INFORMATION HIDING IN COMMUNICATION NETWORKS

T0335596

INFORMATION HIDING IN COMMUNICATION NETWORKS

Fundamentals, Mechanisms, Applications, and Countermeasures

Wojciech Mazurczyk
Steffen Wendzel
Sebastian Zander
Amir Houmansadr
Krzysztof Szczypiorski

IEEE Press Series on
Information & Communication
Networks Security
Stamatios Kartalopoulos, Series Editor

IEEE Press

WILEY

Published by John Wiley & Sons, Inc., Hoboken, New Jersey. All rights reserved
Published simultaneously in Canada.

For general information on our other products and services or for technical support, please contact our Customer Care Department within the United States at (800) 762-2974, outside the United States at (317) 572-3993 or fax (317) 572-4002.

Wiley also publishes its books in a variety of electronic formats. Some content that appears in print may not be available in electronic formats. For more information about Wiley products, visit our web site at www.wiley.com.

Library of Congress Cataloging-in-Publication Data is available.

ISBN: 978-1-118-86169-1

Printed in the United States of America

10 9 8 7 6 5 4 3 2 1

*Wojciech Mazurczyk would like to dedicate this book
to his wife Magdalena and sons Bartek and Tomek.*

Steffen Wendzel would like to dedicate this book to Mali.

*Sebastian Zander would like to dedicate this book
to Wunna, Lara, and Lukas.*

*Amir Houmansadr would like to dedicate this book
to the memory of his grandmother Fatemeh.*

*Krzysztof Szczypiorski would like to dedicate this book
to the memory of his father Jan Szczypiorski.*

CONTENTS IN BRIEF

CONTENTS

LIST OF FIGURES

LIST OF TABLES

FOREWORD

Steganography—the art and science of concealed communication—can be traced back to antiquity. Secret messages written in invisible ink, printed in microdots, or hidden in innocuous hand-crafted images form the history of this exciting field. Systematic research in steganography only began in the late 1990s and early 2000s. Much of this early research focused on hiding data in multimedia content such as digital images, video streams, or audio data and was driven by the quest to protect copyright. At the same time, steganography was seen as a versatile tool to mitigate governmental bans on the use of cryptography. The research performed in these decades gave us a fair understanding of the possibilities and limits of data hiding.

The new hotspot of the field is network steganography. In contrast to many previous approaches that predominantly targeted multimedia data, network steganography attempts to conceal secret messages directly in network streams. It turns out that the ever-increasing volume of Internet traffic provides a perfect cover for steganographic communication. For example, one can utilize unused bits in network protocols to send covert information or change order and timing of network packets to encode supplementary data.

Network steganography has the potential to circumvent oppressive government surveillance by providing means to communicate ''under the radar'' of current network monitoring tools. Steganographic techniques can also avoid censorship by concealing the ultimate goal of a communication channel. Furthermore, techniques similar to those employed in network steganography allow to obfuscate the type of traffic or allow to watermark network flows should be. The goal of the former is to conceal the true purpose of a communication channel, while the latter attempts to trace traffic even if it flows through several networked devices. On the downside, network steganography may be used by attackers to efficiently exfiltrate secrets from highly protected computers or by botnets to set up covert control channels; flow watermarking has the potential to break anonymization tools.

Research in network steganography and related disciplines will give us a good insight into the opportunities and risks of this novel technology, which we just started to explore in detail. We learned that simple steganographic schemes that substitute parts of an ongoing communication with secrets are usually detectable, as they introduce unnatural patterns in data streams. This created opportunities to develop specially crafted steganalytic algorithms that discriminate innocuous from steganographic communication, which in turn led to the development of better steganographic tools. This ''cat-and-mouse'' game between the steganographer and the steganalyst is likely to continue in the near future. The same holds for traffic obfuscation: schemes optimized to mimic a certain distribution of packets will likely be broken with higher order statistics.

I am therefore delighted to see the first comprehensive book on network steganography and related technologies, which I expect will be the standard reference on the subject. I hope that this book will inspire many researchers to explore this exciting discipline of network security—and that it boosts the ''cat-and-mouse'' game between steganographers and steganalysts, which is vital to move our field forward.

STEFAN KATZENBEISSER

PREFACE

Information hiding techniques have their roots in nature, and they have been utilized by humankind for ages. The methods have evolved throughout the ages, but the aims remained the same: hiding secret information to protect them from untrusted parties or to enable covert communication. The latter purpose has grown in importance with the introduction of communication networks where many new possibilities of data hiding emerged.

Information hiding can be utilized for both benign and malicious purposes. Currently, the rising trend among Black Hats is to equip malware with covert communication capabilities for increased stealthiness. On the other hand, covert channels are also becoming increasingly useful for circumventing censorship in oppressive regimes. The complexity and richness of continuously appearing new services and protocols guarantee that there will be a lot of new opportunities to hide secret data. A problematic aspect in this regard is the lack of effective and universal countermeasures that can be applied in practice against increasingly sophisticated information hiding techniques (especially when used for malicious purposes).

Security, censorship, and blocking are on the rise in the Internet. Hence, where covert communication techniques seemed like overkill some time ago, they may become very attractive in the future. Therefore, we expect that in the future, information hiding methods for communication networks will see more widespread use than today, and they will continue to become more sophisticated and harder to detect. It must be emphasized that the threat posed by information hiding techniques can potentially affect every Internet user, since even innocent users' network traffic can be utilized for covert communication purposes (without their explicit knowledge). This will raise similar legal and ethical issues like we are currently experiencing with botnets.

APPROACH AND SCOPE

We decided to write this book, because there was no reference book available that covers all aspects of information hiding for communication networks from the history, over the hiding techniques, to the countermeasures. We formed a team of authors, each with significant expertise in certain areas of the overall topic, who contributed equally to the book. As a group, we were able to put together a comprehensive description of the current state-of-the-art of information hiding in communication networks, including the important issues, challenges, emerging trends, and applications.

This book is intended to be utilized mainly as a reference book to teach courses like information hiding, or as a part of network security or other security-related courses. The target audience of the book are graduate students, academics, professionals, and researchers working in the fields of security, networking, and communications. However, the first few chapters of this book are written so that non-expert readers will be able to easily grasp some of the fundamental concepts in this area.

The book is divided into eight chapters that cover the most important aspects of information hiding techniques for communication networks. The last chapter concludes this book.

CHAPTER OVERVIEW

Chapter 1 is written mostly in a tutorial style so that even a general reader will be able to easily grasp the basic concepts of information hiding, their evolution throughout the history, and their importance especially when utilized in networking environments. It also contains many examples of applications of modern information hiding for criminal and legitimate purposes, and it highlights current development trends and potential future directions.

Chapter 2 discusses the existing terminology and its evolution in the information hiding field. It introduces a new classification of data hiding techniques; however, our new classification builds on existing concepts. The chapter then introduces the two main subfields: network steganography and traffic type obfuscation methods. The chapter concludes with a description of the model for hidden communication and related communication scenarios. It also highlights potential countermeasures.

Chapter 3 describes in detail different flavors of network steganography. Three main types of techniques are distinguished and then characterized: hiding information in protocol modifications, in the timing of network protocols, and hybrid methods.

Chapter 4 introduces techniques that improve the resiliency and undetectability of network steganography methods. These techniques are usually implemented by so-called control protocols. The chapter discusses their features, highlights the design of known control protocols, and discusses control protocol-specific engineering methods.

Chapter 5 concentrates on traffic type obfuscation techniques that allow to hide the type of the network traffic exchanged between two (or multiple) network entities, that is, the underlying network protocol. Typical applications of these methods are twofold: blocking resistance or privacy protection. The chapter presents a classification of traffic type obfuscation techniques and covers the most important of these techniques in detail.

Chapter 6 focuses on network flow watermarking. Network flow watermarking manipulates the traffic patterns of a network flow, for example, the packet timings, or packet sizes, in order to inject an artificial signal into that network flow—a watermark. This watermark is primarily used for linking network flows in application scenarios where packet contents are striped of all linking information.

Chapter 7 presents most recent examples and applications of information hiding in communication networks with a focus on current covert communication methods for popular Internet services. This includes hiding information in virtual worlds (e.g., multiplayer online games), IP telephony, wireless networks and modern mobile devices, and P2P networks and their global services like BitTorrent and Skype. Additionally, we discuss potential steganographic methods for social networks and the Internet of Things (e.g., building automation systems).

Chapter 8 discusses potential countermeasures against network steganography. The chapter describes different types of techniques that lead to the detection, prevention, and limitation of hidden communication.

Chapter 9 concludes the book.

WOJCIECH MAZURCZYK

STEFFEN WENDZEL

SEBASTIAN ZANDER

AMIR HOUMANSADR

KRZYSZTOF SZCZYPIORSKI

ACKNOWLEDGMENTS

Wojciech Mazurczyk would like to thank his family for their love, encouragement, and continuous support. He is also grateful to all colleagues and co-workers with whom it was an honor to collaborate and who have contributed to the research presented in this book.

Steffen Wendzel would like to thank all his co-authors of the last years and Jaspreet Kaur for her contribution of aspects on countermeasures against steganographic control protocols.

Sebastian Zander would like to thank Grenville Armitage, Philip Branch, and Steven Murdoch for the fruitful collaborations and their contributions to some of the research presented in this book. Sebastian would also like to thank his family for their constant encouragement and support.

Amir Houmansadr would like to thank his wife, Saloumeh, for her immense support, his son, Ilya, for bringing joy to their lives, and the rest of his family for their love. He would also like to thank all of his collaborators who have contributed to the research presented in this book, including Nikita Borisov, Negar Kiyavash, and Vitaly Shmatikov.

Krzysztof Szczypiorski would like to thank Wojciech Mazurczyk, Józef Lubacz, Piotr Białczak, Krzysztof Cabaj, Roman Dygnarowicz, Wojciech Frączek, Iwona Grabska, Szymon Grabski, Marcin Gregorczyk, Bartosz Jankowski, Artur Janicki, Maciej Karaś, Bartosz Lipiński, Piotr Kopiczko, Paweł Radziszewski, Elżbieta Rzeszutko, Miłosz Smolarczyk, Paweł Szaga, and Piotr Szafran for fruitful cooperation in the area of network steganography in the last 12 years.

ACRONYMS

AAL	Ambient Assisted Living
AH	Authentication Header
AODV	Ad Hoc On-Demand Distance Vector
API	Application Programming Interface
APT	Advanced Persistent Threat
ARQ	Automatic Repeat Request
BACnet	Building Automation and Control Networking Protocol
BYOD	Bring Your Own Device
C&C	Command and Control
CCE	Corrected Conditional Entropy
CCN	Content-Centric Networks
CE	Conditional Entropy
CFG	Context-Free Grammar
CFT	Covert Flow Tree
CFTP	Covert File Transfer Protocol
CRC	Cyclic Redundancy Check
CSLIP	Compressed Serial Line Interface Protocol
CSMA/CD	Carrier Sense Multiple Access/Collision Detection
CT	Covert Transmission
CTS	Clear to Send
DCT	Discrete Cosine Transform
DDC	Direct Digital Control
DF	Don't Fragment
DHCP	Dynamic Host Configuration Protocol
DHT	Deep Hiding Techniques
DHT	Distributed Hash Table
DLP	Data Leakage Protection
DNS	Domain Name System
DoD	Department of Defense
DPI	Deep-Packet Inspection

DRM	Digital Rights Management
DSP	Digital Signal Processor
DSSS	Direct Sequence Spread Spectrum
DTS	Direct Target Sampling
DWT	Discrete Wavelet Transform
ECG	Electrocardiogram
ESP	Encapsulated Security Payload
FCFS	First Come First Serve
FCS	Frame Check Sequence
FPE	Format-Preserving Encryption
FPGA	Field-Programmable Gate Array
FPSCC	FPS Covert Channel
FPS	First Person Shooter
FR/R	Fast Retransmit and Recovery
FTE	Format Transforming Encryption
FTP	File Transfer Protocol
GMM	Gaussian Mixture Models
GPS	Global Positioning System
GUI	Graphical User Interfaces
HTML	HyperText Markup Language
ICMP	Internet Control Message Protocol
ICS	Industrial Control System
IH	Information Hiding
IoT	Internet of Things
IP	Internet Protocol, version 4 (also IPv4)
IPD	Interpacket Delay
IPS	Inter Protocol Steganography or Intrusion Prevention System
IPSec	IP Security
IPv6	Internet Protocol, version 6
IRC	Internet Relay Chat
ISN	Initial Sequence Number
ISO	International Organization for Standardization
ISP	Internet Service Provider
JPEG	Joint Photographic Experts Group
LACK	Lost Audio Steganography
LAN	Local Area Network
LSB	Least Significant Bit
LTE	Long-Term Evolution

MAC	Medium Access Control
MFCC	Mel-Frequency Cepstral Coefficients
MITM	Man-in-the-Middle
ML	Machine Learning
MLS	Multilevel Security
MOS	Mean Opinion Score
MPEG	Motion Picture Experts Group
MSE	Mean Squared Error
MS/TP	Master–Slave/Token Passing
MTU	Maximum Transmission Unit
NAAW	Network-Aware Active Warden
NAT	Network Address Translation
NEL	Network Environment Learning
NOOP	No Operation
NTP	Network Time Protocol
OFDM	Orthogonal Frequency-Division Multiplexing
OLSR	Optimized Link-State Routing
ON	Ordinary Nodes
OS	Operating System
OSI	Open Systems Interconnection
OSN	Online Social Network
OT	Overt Transmission
P2P	Peer to Peer
PC	Protocol Channel
PCAW	Protocol Channel-Aware Active Warden
PDF	Portable Document Format
PDU	Protocol Data Unit
PEX	Peer Exchange
PHCC	Protocol Hopping Covert Channel
PLC	Packet Loss Concealment
PLL	Phase Lock Loop
PLPMTUD	Packetization Layer Path MTU Discovery
PMTUD	Path MTU Discovery
PSCC	Protocol Switching Covert Channel
PSDU	Physical Layer Service Data Unit
PSNR	Peak Signal-to-Noise Ratio
PT	Payload Type
QoC	Quality of Covertness

QoS	Quality of Service
RFC	Request for Comments
RSTEG	Retransmission Steganography
RTCP	Real-Time Transport Control Protocol
RTO	Retransmission Timeouts
RTP	Real-Time Transport Protocol
RTS	Request to Send
RTT	Round-Trip Time
SACK	Selective Acknowledgment
SAFP	Store and Forward Protocol
SBC	Session Border Controller
SCCT	Smart Covert Channel Tool
SCTP	Stream Control Transmission Protocol
SDP	Session Description Protocol
SGH	Steganogram Hopping
SIP	Session Initiation Protocol
SkyDe	Skype Hide
SN	Super Nodes
SOHO	Small Office Home Office
SoM	Start of Message
SR	Secret Receiver
SRM	Shared Resource Matrix
SS	Secret Sender
SSH	Secure Shell
SVM	Support Vector Machine
TCP	Transmission Control Protocol
TLS	Transport Layer Protocol
ToS	Type of Service
ToU	Type of Update
TranSteg	Transcoding Steganography
TrustMAS	Trusted Multiagent System
TTL	Time to Live
TTO	Traffic Type Obfuscation
UDP	User Datagram Protocol
UGS	Unsolicited Grant Service
UMTS	Universal Mobile Telecommunications System
USB	Universal Serial Bus
VoIP	Voice over IP

VPN	Virtual Private Network
VSC	Virtual Sound Card
WEP	Wired Equivalent Privacy
WiMAX	Worldwide Interoperability for Microwave Access
WiPad	Wireless Padding
WLAN	Wireless Local Area Network

1

INTRODUCTION

If you want to keep a secret, you must also hide it from yourself.

—George Orwell, 1984

Modern information hiding techniques for communication networks typically conceal data in network traffic by utilizing various characteristic features of network protocols, for example, their control information, behavior, or relationships with other protocols. However, it must be noted that the essence of these methods is not some recent invention. In the following, we will review the history, basics, and then current trends of the information hiding methods.

1.1 INFORMATION HIDING INSPIRED BY NATURE

Information hiding techniques that conceal the presence of secret data from curious third party observers have proved very handy and have been utilized and mastered by humankind throughout the ages. However, it must be noted that the inspiration for such

Information Hiding in Communication Networks: Fundamentals, Mechanisms, Applications, and Countermeasures,
First Edition. Wojciech Mazurczyk, Steffen Wendzel, Sebastian Zander, Amir Houmansadr, and Krzysztof Szczypiorski.
© 2016 by The Institute of Electrical and Electronics Engineers, Inc. Published 2016 by John Wiley & Sons, Inc.

mechanisms is strongly related to phenomena observable in the kingdoms of living things.

Clearly nature, with its about 3 billion years of experience in evolution by natural selection, genetic drift, and mutations and the magnitude of species, is the most amazing and recognized invention machine on Earth. Thus, it is not surprising that it has inspired inventors and researchers for ages. Nature's footprint is present in the world of information technology, where there are an astounding number of bio-inspired computational techniques and networking technologies. Examples include genetic algorithms, neural networks, artificial immune systems, or sensor networks—just to name a few. Similar replicas from nature can also be found in the field of digital security where defensive and offensive strategies happen to mimic the ongoing species' race of arms [1].

An analogous situation can be observed in the information hiding field. Evolution has proved long ago that abilities to disguise can serve as a perfect protection and can significantly improve chances for survival. In nature, the obligatory rule has been always "fight or hide," meaning that only superior predators and skilled prey are able to prevail, especially as resources that are needed for survival are very often limited and competed for.

Hence, an organism requires competitive dominance to survive—this can be achieved, for instance, by masterfully blending into the surrounding environment. Therefore, let us take a closer look at what information in nature can be subjected to hiding and how it relates to the modern information hiding techniques.

In nature, the most obvious way to mask an animal's presence from external threats is to physically hide in some kind of shelter such as a hole or hollow, to make itself hard to notice and/or hide beyond the reach of a predator. However, there are many more subtle and sophisticated possibilities. In ecology, the term *crypsis* was coined [2] to describe the abilities of organisms to effectively hide or conceal their presence in order to avoid detection/observation. The two most notable crypsis techniques are *camouflage* and *mimicry*.

Camouflage embraces all solutions that utilize physical shape, texture, coloration, illumination, and so on in making animals hard to spot. This causes the *information about their exact location to remain ambiguous*. Organisms with camouflage typically look like an element of their habitat, for example, a rock, twig, or leaf. Examples of living things that can easily blend into the background include a chameleon that can make itself look similar to a leaf or a jellyfish that is almost transparent to hide itself in water.

However, patterns and/or colorations can also be used to confuse the predator, that is to make *information about the prey hard to interpret*. Such, so-called "disruptive," camouflage is possible and can be seen in, for example, a herd of zebras, where it is difficult for an attacking lion to identify a single animal in a herd when they flee in panic.

Mimicry characterizes all the cases in which organism's attributes are obfuscated by adopting the characteristics of another living organism. In particular, this means that the prey can avoid an attack by making the predator believe it is something else; for example, a harmless species can mimic a dangerous one. Therefore, in fact what the

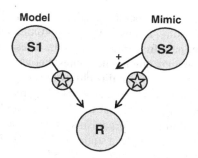

Figure 1.1. Basic mimicry system. S1 and S2 denote signal transmitters and R is the signal receiver. "+" denotes that the response of the receiver R is advantageous to S2; thus, S2 benefits from the S1/R couple. (Reproduced from [3] with permission of Wiley.)

prey accomplishes is that it hides *information about its own identity* by impersonating something that it is not. For example, a wasp moth mimics the coloring of the real, stinging wasp. It is its protective measure because the predators are less likely to attack if they can be potentially harmed. A similar situation happens for harmless milk snakes that bear significant resemblance to venomous coral snakes.

The basic mimicry model, described in [3], is illustrated in Figure 1.1. Its main components include two different signal transmitters (denoted as S1 and S2) that transmit similar signals to the signal receiver (R). S1 is called a *model* and S2 that produces a "counterfeit" signal for the receiver is called a *mimic*. R reacts similarly to both S1 and S2 because it is unable to distinguish between the two received signals. This means that mimicry occurs when the receiver is deceived by the signal transmitted from S2 and interprets it as if it was originally generated by S1. In other words, S2 spoofs S1's identity by exploiting R senses' imperfectness. Typically, the response of the receiver R is advantageous (the "+" sign in Figure 1.1) to the mimic S2 in the sense that if R was not deceived then it would pose a threat or cause harm to S2.

It must be emphasized that the above-described mimicry model bears, in principle, many resemblances with information hiding. This includes the following similarities:

- there is a signal transmitter that wants to keep a secret uncovered;
- some technique is utilized to hide a secret;
- hiding of the secret raises no suspicion when inspected by a third-party observer;
- the third-party observer is deceived due to its inability to detect with certainty where the secret is located;
- the signal transmitter benefits from staying undetected.

But not only the information about an organism's presence or identity can be subject to hiding. A precious evolutionary achievement, which can be used for both defensive and offensive purposes, is an ability to hide the *information about the communication process* between the organisms, for example, by utilizing a covert communication channel. The value of this skill has proven high among the Philippine tarsiers

(*Tarsius syrichta*), which are small nocturnal primates. It was discovered that they have a high-frequency limit of auditory sensitivity of approximately 91 kHz and are also able to vocalize with a dominant frequency of 70 kHz. These values are an example of ultrasonic communication and are among the highest known for terrestrial mammals. It is believed that Philippine tarsiers utilize this ultrasonic communication as their private covert communication channel that is undetectable by predators, prey, and potential competitors [4]. It must be emphasized that this is not a stand-alone case for utilizing hidden communication capabilities by organisms. In fact, there are many examples illustrating this phenomenon; for example, a squid (*Loligo pealeii*) and a cuttlefish (*Sepia officinalis* L.) produce polarized patterns on their bodies that are thought to be used as a covert communication channel, invisible to animals lacking polarization vision [5].

The summary of analogies between nature and information hiding techniques in communication networks is presented in Table 1.1.

Relationships similar to nature's predator–prey association can also be discovered for information hiding methods and their countermeasures. As mentioned earlier, in nature there is always a predator–prey association of some kind involved. Where conflict of interest is present, the reason behind it is either gaining competitive dominance over counterparts or obtaining access to a limited resource. This leads to a situation where known interaction models are constantly adapted to devise new means of outwitting the rival [1].

The ongoing evolution of both offensive and defensive techniques in nature finds analogy in information hiding methods developed by humans and their countermeasures. Indeed, in both cases we are witnessing an arms race, however taking place in different time windows. The general rule is that for every offensive technique that has been developed, sooner or later, a defensive scheme appears. An example from the natural environment are mollusks, such as *Murex* snails, which have developed thick shells to avoid being eaten by animals such as crabs and fish. In turn, predators such as crabs have grown more powerful claws and jaws that compensate for the snails' thick shells. Similarly, if a new information hiding method is developed, then, typically, sooner or later a dedicated mitigation technique appears. In the next step, to avoid further disclosure, the hiding method is "upgraded" but only to a point that just makes it

T A B L E 1.1. Analogies between the information hiding field and the kingdoms of living things.

Subject of Concealment	Kingdom of Living Things	Information Hiding Technique
Location/presence	Shelter, camouflage techniques	Steganography
Information meaning	Mimicry, disruptive camouflage	Cryptography
True identity or intentions	Mimicry	Anonymity and traffic type obfuscation techniques
Communication process	Species' covert communication	Network steganography

invisible again; that is, typically, the lowest investment to achieve the minimum desired goal is utilized. This arms race in both worlds (real and virtual) seems to be never ending.

The above-mentioned astounding examples from nature prove that various information hiding techniques and their countermeasures invented by humans have been significantly bio-inspired.

1.2 INFORMATION HIDING BASICS

From the historical point of view, the oldest two terms related to the information hiding are steganography and cryptography. They originate from the ancient Greek words steganos, meaning protected (covered), and kryptos, meaning hidden (secret). Although the Greek terms are semantically quite close, it is plausible to consider steganography and cryptography as different methods of hiding information: steganographic methods hide information, thereby making it "difficult to notice" (by means of embedding it in an information carrier), while cryptographic methods hide information by making it "difficult to recognize" (by means of transforming it).

Both cryptography and steganography techniques are practically applied in imperfect communication environments imposed by physical features of information carriers. While this imperfectness is generally an obstacle for cryptography, it is an essential enabling condition for many steganographic techniques that utilize redundant communication mechanisms (protocols) implemented to cope with such imperfect environments (to provide reliable communication). In principle, a message to be hidden with the use of a steganographic technique may be first encrypted with some cryptographic technique. Note, however, that if a message is encrypted, this may increase the probability that the message is noticed and thus reduces the chance of achieving the principal goal of the steganographic method in use. It should also be stressed that without encrypting the message, steganography provides only *security by obscurity* and is therefore violating Kerckhoff's principle (which demands that the cryptosystem itself should be public and the secrecy should lie only in a secret key).

As will be explained in the following, there is a trade-off between the effectiveness of any steganographic technique (in terms of potential steganographic capacity) and its susceptibility to steganalysis (i.e., to being uncovered). Essentially, evaluating the effectiveness of a particular technique requires analysis of its robustness to detection. This, in principle, requires considering potential methods of uncovering the hidden communication, which constitutes another distinctive feature of steganography with respect to cryptography. Considering the above fact, the main objectives, features, and potential applications of information hiding with the use of steganographic and cryptographic methods should not be regarded as competitive and/or complementary.

In general, information hiding includes within its scope various techniques that can be broadly divided, based on the aim to be achieved, into two groups:

- *Solutions that allow to hide secret data in such a way that no one besides the owner is authorized to discover its location and retrieve it.* In other words, the aim is to not reveal the secret to any unauthorized party. A good analogy here is a locker with confidential documents, which is placed somewhere in the house and its location is known only to its owner. He/she is able to retrieve his/her secrets and can influence the content of the locker at any time he/she wants. However, someone who is unaware of the locker location is not able to access it.

- *Methods intended for the communication of messages with the aim of keeping some aspect of such exchange secret.* Typically, three aspects of such transmission can be subjected to concealment: *(i)* identity of transmitting and/or receiving party; *(ii)* communication content; and *(iii)* communication process, that is, the fact that the secret message exchange takes place. A straightforward example is a "live drop," that is, clandestine communication between two spies on hostile territory. When they want to exchange confidential information, first they must signal each other that a meeting is necessary. At the meeting point, they both want to "blend" into the background so that they will not be recognized easily. Then they utilize various masking and/or obfuscation techniques to conceal the fact that their communication is taking place and secrets are exchanged.

Throughout the ages, humans utilized both forms of information concealment. With respect to modern techniques, examples of the first group of information hiding methods include digital media steganography, that is, embedding secret data into innocent-looking digital images, audio files, or video files just to conceal their existence and the data can be retrieved later if needed. Analogously, there were also historical methods that functioned in the same way by concealing secrets in paintings (for details on the method, please see the next section).

The latter case, where some form of communication is involved, is obviously closer to the scope of this book as the same aspects can be subjected to concealment in communication networks. This incorporates, among others,

- *techniques for anonymity*—where identities of communicating parties are to remain secret;
- *cryptography*—to hide the true meaning of the transmitted content;
- *communication hiding methods*—to disguise the existence of the communication process.

A simple example representing the latter case, that is, communication hiding methods, involves a query response type of exchange of messages for which the communication protocol assumes that the response should come within a specific time limit; otherwise, it is treated as excessively delayed and discarded. Communicating parties that want to use this protocol for information hiding purposes make an agreement, which becomes their shared secret, that the responses carrying hidden information will be purposefully excessively delayed and that such responses will be read by the recipient (i.e., not discarded). This "trick," that is, manipulation of the communication protocol, may be effective only if the communication channel introduces some variable delay in the message transmission. Potential observers of the communication (potential attackers), who know the communication protocol and follow the message exchange, do not become suspicious of the existence of hidden communication if the frequency of excessively delayed responses is not considered to be abnormal, that is, does not exceed some expected frequency that the observers assume, based on their knowledge of delay properties of the communication network.

With regard to historical techniques, one of the first and most well-known historical examples of concealing the communication process is the passing of information by means of strategically placed tattoos. The invention of this technique is attributed to the Greek tyrant of Miletus—Histiaeus—who lived in the late sixth century BC. More specifically, he had chosen his trusted slave and commanded him to tattoo a secret message onto his shaved head. Then he waited for the hair to regrow to cover the tattoo, and sent the messenger to the intended recipient, who shaved off the hair to read the secret instructions to initiate a revolt against the Persians.

It is worth noting that this historical example bears many resemblances with the current information hiding techniques utilized in communication networks (Figure 1.2), such as the intentional delays-based method mentioned earlier. These resemblances are summarized in Table 1.2.

First of all, the most obvious similarity is the need for communication between the secret data sender and the receiver that cannot be revealed. Hence, some form of information concealment method is required to form a so-called *covert channel* between sender and receiver. In the ancient method, tattoos on human skin were used so that the secret data would escape the attention of guards stationed on the country's roads. The modern technique relies on introducing additional delay to the existing datagrams of a network flow with the main aim being to deceive other users/devices in the network.

Moreover, in both cases a selection of an appropriate *carrier* is vital. Generally, the most suitable carrier for information hiding purposes must fulfill two requirements: *(i)* it must be commonly used so that its existence is not considered as an anomaly itself; and *(ii)* the carrier modification caused by the embedding of the secret data must not be ''noticeable'' to anyone.

Both these requirements are fulfilled for the historical method. In ancient times, the messengers were a popular way of communication over distances and, moreover,

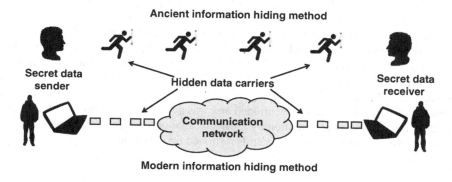

Figure 1.2. Ancient and modern information hiding.

T A B L E 1.2. Analogies between exemplary ancient and modern information hiding techniques.

Method	Ancient	Modern
Main aim	To conceal the fact of communication and the exchanged data	
Relies on	Strategically placed tattoos	Excessively delayed datagrams
Carrier used	Human skin	Network flow
Carrier popular	Yes	Yes, if a flow is properly selected
Modification of the carrier	Unnoticeable	

the servant could carry also some phony correspondence (an overt message) as a decoy to deceive potential adversaries. Moreover, after the hair grew back the tattoo was concealed, so the confidential instructions were not easy to spot for people unaware where to seek for them. Similarly, both requirements are fulfilled for the modern method mentioned earlier that relies on introducing excessive delays. First, the traffic flow selected for the information hiding purposes must be related to some common service or protocol that can be typically encountered in the network. Second, due to the fact that additional delays are a natural phenomenon in IP networks, adding a reasonably small delay should not raise any suspicion.

Many more historical examples of how humankind utilized information hiding techniques will be covered in the following section.

1.3 INFORMATION HIDING THROUGHOUT THE HISTORY[1]

As mentioned earlier, the roots of the utilization of information hiding by humankind stem from antique times. The ancient Greeks, who sought inspiration in nature, had considered the ability to simulate its ways as a measure of craftsmanship. Inherently, the ancient people picked ordinary objects as a carrier for the secret message. The physical object, possibly even a living organism, had to be transported from one participant of the communication to the other without raising suspicion on the way.

Therefore, it is not surprising that the first written report on the use of information hiding is attributed to the Greek historian, Herodotus. The reported method involved camouflaging a secret message within a hare corpse [7]. The animal was meant to imitate a game trophy and was carried by a man disguised as a huntsman. In this way, a message could be passed without raising unnecessary suspicion. Another steganographic technique that has been already mentioned and was noted in Herodotus's Histories was based on the passing of information by means of a tattoo on a servant's head, covered with hair to escape the attention of guards stationed on the country's roads.

---

[1] The following section is written based on the publication "Trends in steganography" by E. Zielińska, W. Mazurczyk, and K. Szczypiorski [6].

The most notable method quoted in the historian's works is the communication on wooden tablets—these were usually coated with a thin layer of wax, on which text would be embossed. Clandestine passing of information with the aid of this medium could be achieved if the text was carved permanently on the wood (the carrier of the secret data), and then coated with wax. The tablet would then be passed as an unused tablet, and only an aware recipient would know that the letters would become visible if the wax coat was melted.

The Greek methods were fairly easy to implement as they relied on common patterns—the messages that were passed utilized a carrier cover that could be considered common at the time. Alongside the progress of human civilization and the way people communicated, new opportunities arose. The popularization of parchment, which substituted papyrus, brought about a new cover for steganograms. Its popularity led to the development of complementary steganographic algorithms, capable of exploiting the new cover's properties. Pliny the Elder is considered to be the inventor of sympathetic inks [8], as he postulated the use of tithymalus plant's sap to write text, which would become invisible upon drying. A subtle heating process would lead to the charring of the organic substances contained in the ink, which would then turn brown and make the text visible.

The common factor of all of the aforementioned techniques is the operation of adding surplus content (additional features) to a carrier, which otherwise would not physically contain the inserted elements.

A different type of steganography that was invented in ancient Rome is the semagram, or a secret message that does not take written form. Tacitus, the historiographer of the ancient world, became interested in the astragali [9], which were small dice made of bone. Such objects could be threaded onto a string, where the placement of holes could encode a secret message. A properly crafted object would pass unnoticed as a toy.

The Medieval Ages had brought about major progress in the art of information hiding. The Chinese invention of paper, upon its introduction to Europe, in the Middle Ages, had brought forth the necessity of differentiating between manufacturers' products. This is how paper watermarking was born [10]. Presently, digital watermarking and digital image steganography, stemming from the mentioned invention, are based on the same principle. File watermarking is now considered a separate branch of the information hiding techniques. Petitcolas, Anderson, and Kuhn, in their survey paper from 1999 [11], have derived a whole field of copyright marking, of which watermarking is a subclass. The current notion refrains from classifying digital watermarking as steganography, due to the lack of an explicit communication aspect and the fact that for embedded watermarks robustness is more important than providing "invisibility" to the participants of communication.

The popularization of paper had further consequences. The steganographic vector was no longer necessarily a physical object, but could take written form, where the carrier text itself would conceal the privileged information. Among the inventions that achieved popularity during the Medieval times are the textual steganographic methods, including among others, the *acrostic*. This term refers to pieces of writing, whose first letters or syllables spell out a message. The most famous example of such textual steganography is attributed to a Dominican priest named Francesco Colonna, who in 1499 had hid in his book "Hypnerotomachia Poliphili" a love confession that could be spelled out from the first letters of subsequent chapters [12].

A more sublime carrier is the language itself, as the Medieval people had discovered. Here, the embedding process occurs in the linguistic syntax and semantics. Linguistic steganography may be derived from the aforementioned technique of textual steganography, as it relies on the manipulations of the written (possibly even spoken) language with the aim of tricking the perception of an unaware dupe. Following the postulates of Richard Bergmair [13], linguistic steganography covers within its scope any technique that involves intentional mimicry of typical structures of words characteristic to a specific language. This may concern the deliberate tampering with grammar, syntax, and the semantics of a natural language. Any action involving modification of the aforementioned aspects should maintain the innocent appearance of the cover text.

The Renaissance had brought about an invention by an Italian-born scientist Giambattista della Porta who, in the 16th century, detailed how to hide a message inside a hard-boiled egg: write on the shell using ink made from a mixture of alum[2] and vinegar. The solution penetrated the eggshell, leaving no trace on the surface, but a discoloration occurred on the white, leaving the message on its surface, which was only readable once the shell was removed.

The famous Vexierbild (puzzle picture), in the form of woodcut, was created by Erhard Schön, a Nürnberg engraver in the early years of the 16th century. When looking at it normally, one would see a nightmare landscape, but once viewed from a proper angle the picture revealed portraits of famous kings. The message hidden in the woodcut had not been crafted to entertain the viewer, but to hide its political message from those that Schön did not want to see it.

Gaspar Schott, a German Jesuit from the Age of Enlightenment, followed the trail marked by his Renaissance predecessors. His work, published in 1680, entitled "Schola Steganographica," explained how to utilize music scores as a hidden data carrier. Each note corresponded to a letter, which appeared innocent as long as nobody attempted to play the odd-sounding melodies.

The Industrial Revolution, which followed the Age of Enlightenment, had brought about new means of communication. Newspapers became a popular and reliable source of the latest information. At some point, it became obvious that a newspaper could serve as a perfect steganographic carrier. Since daily papers could be sent free of charge, it was convenient to poke holes over selected letters and thus craft a secret message. This is how "newspaper codes" were born.

The first symptoms of the growing interest in steganography may be traced back to the period of the two World Wars and then the Cold War. These had brought about steganographic techniques such as microdots—punctuation marks with inserted microscopic negatives of images or texts [14].

---

[2] A colorless soluble hydrated double sulfate of aluminum and potassium used in medicine as a styptic and astringent, in the manufacture of mordants and pigments, and in dressing leather and sizing paper.

The period of World Wars was a true bonanza of hidden communication schemes. The First World War was witness to the spectacular return of all sorts of invisible inks [15]. The Second World War was marked by Hedy Lamarr's and George Antheil's patent for spread spectrum communication [16]. They devised a method for guiding torpedoes with a special, multifrequency set of signals, resistant to jamming attempts. The control information was dispersed over a wide-frequency bandwidth, which provided cover. The idea of embedding information in a number of different frequencies later found use in the fields of digital image and audio steganography. Most of the described historical methods are presently no longer in use, but as noted, some of them have given birth to modern information hiding techniques. Many of the currently exploited digital techniques have their analogue predecessors, as is the case with digital watermarking, which derives from paper watermarking.

The technological development in the 20th century had also accelerated the development of more sophisticated techniques. Among these inventions were the so-called ''subliminal channels'', which are based on the embedding of steganograms in cryptographic protocols. The main principle was to insert content into digital signatures. This concept was introduced in 1984 by Gustavus Simmons, despite the U.S. government's prohibition on publishing of materials on steganography. Simmons proposed that overt and monitored communication conducted between two participants can be supplemented with a steganographic channel. This channel would be based on a number of dedicated bits of the message authentication. These, at the cost of reducing the message authentication capability of the digital signature, would serve as the steganographic channel capacity [17]. Subliminal channels utilized the cryptographic protocol as the carrier for steganograms.

The fall of the 20th century was also marked by the discovery of a steganographic technique that succeeded to proliferate most of the globe without raising suspicion. Apparently, a number of printer manufacturers introduced to the global market a series of printers enriched with the capability to trace back the originating device of a printout [18]. The intention was to enable forensic investigation in case it was needed to track down the source of a document.

1.4 EVOLUTION OF MODERN INFORMATION HIDING

From the survey of historical methods presented in the previous section, it is clear that primarily information hiding methods were designed to deceive human senses such as vision, hearing, or touch. However, because such techniques evolved alongside the evolution of the human communication possibilities, the scope broadened. Currently, modern information hiding methods utilize the 20th century's inventions—computers and networking. In communication networks, a human end user is not only the one to be potentially deceived. More often, information hiding techniques must also succeed in tricking networking systems and/or devices to avoid disclosure. The evolution of information hiding that involves historical as well as modern methods is summarized in Figure 1.3.

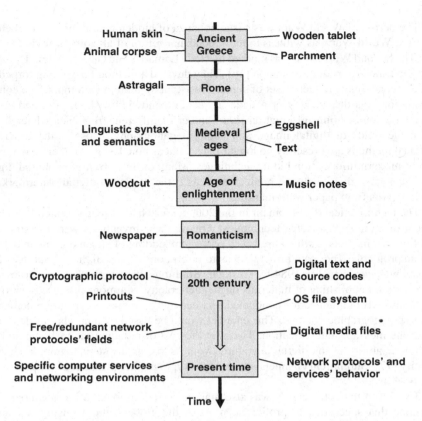

<u>Figure 1.3.</u> Evolution of hidden data carrier throughout the history. (Reproduced from [6] with permission of ACM.)

The following two main trends of development of the so-called digital information hiding techniques can be distinguished:

- *Digital steganography*, which includes techniques for digital media content, linguistic, file system, and network steganography.
- *Traffic type obfuscation* methods.

These two main branches of digital information hiding are explained and described in the following sections. It must also be emphasized that currently the majority of research in these areas is devoted to data hiding in digital media, network steganography, and traffic type obfuscation methods. Data hiding in digital media is the most mature area with significant achievements; thus, the exploration in this field is presently not as dynamic as in the more recently sprouted fields of network steganography and traffic type obfuscation.

1.4.1 Information Hiding in Digital Content

Digital media steganography dates back to the 1970s, when researchers focused on developing methods to secretly embed a signature in a digital picture. Many different methods were proposed, including, among others, patchwork, least significant bit modifications, or texture block coding [19]. The introduced techniques were intended for images with both lossy compression, for example, JPEG, and lossless compression, for example, BMP.

The variety of algorithms for the embedding of secrets in digital pictures can be grouped according to the type of alterations that were induced. Following Johnson and Jajodia [18], the modifications either are bit-wise, influencing the spatial domain characteristics of the image, or affect the frequency domain characteristics. Alternatively, specific file format intricacies may be exploited. Moreover, a mix of all these techniques is possible. The transform domain which includes all techniques that involve manipulation of algorithms and image transforms provides for the most versatile medium of embedding. Affecting of the image processing algorithms may involve, among others, discrete cosine transform (DCT), discrete wavelet transform (DWT), and Fourier transform, which may result in alterations of, for example, luminance or other measurable properties of an image [18].

Digital image steganography's position is unfaltering as it is the most prominent branch of steganography—the survey paper by Cheddad et al. [20] points out that lately the interest concentrated on employing digital media steganography and watermarking for embedding of confidential, patient-related information in medical imagery. Another application of digital image steganography foreseen to become popular is the implantation of additional data in printed matter, which, invisible for the naked eye, becomes decodable when photographed and processed by a cellular phone [20].

Notably, digital image steganography is mostly oriented toward tricking the human visual system into believing that the perception of the image has not been manipulated in any way. The same rule applies to the whole field of digital media steganography, whose primary function is to trick the observer to believe that the crafted "forgery" is indeed genuine. The "communication aspect" of the steganographic algorithm is secondary to the process of embedding of the secret data.

The human auditory system is equally prone to delusion as the visual perception. The research focus moved to audio files such as MPEGs. The developed techniques included, among others, frequency masking, echo hiding, phase coding, patchwork, and spread spectrum. It also became apparent that error correction coding is a good supplemental carrier for audio steganography—any redundant data can be used to convey the steganogram at the cost of losing some robustness to random errors [19]. This idea later found use in network protocol-based steganography.

Next, steganographers took video files as target carrier. Most of the proposed methods were adaptations of the algorithms proposed for audio and image files. Video-specific solutions involved using either video's I-frames' color space [21] as a steganographic carrier or P-frames' and B-frames' motion vectors [22]. Currently, steganography in video files either takes advantage of the existing methods for audio and image files as mentioned above or makes use of the intrinsic properties of the video transmission, such as movement encoding.

Parallel to digital image and audio steganography, information hiding in text was developed—the available methods exploited various aspects of the written word. The first set of techniques altered word spacing, which was even claimed to have been used at the times of Margaret Thatcher to track leakages of cabinet documents [23]. More advanced steganographic methods used syntactic and semantic structure of the text as a carrier. The introduced methods included displacement of punctuation marks, word order, or alterations of the choice of synonyms, which could be attributed certain meaning.

Presently, some suggest that even email SPAM messages may be a carrier of steganography, due to the large numbers of spam emails that are emitted every day [24]. According to the work by Bennet [25], the possible techniques can either rely on the generation of text with a cohesive linguistic structure or use natural language text as a carrier. It should be pointed out that the first technique does not entirely fulfill the definition of steganography, where the existence of the carrier should be independent of the existence of the injected hidden content. Thus, a text lacking rhetorical structure cannot be considered a proper carrier. Specialists also differentiate between textual steganography and linguistic steganography [25]. The "SPAM method" is a linguistic method, and the embedding occurs with the aid of context-free grammars (CFGs). CFGs have a tree structure; thus, the selection of proper words, or branches, provides encoding for binary data. An example of a textual method would be a substitution technique, where a message's carrier is the set of white spaces and punctuation marks undergoing shifting, repetition, or other modifications.

Parallel to this research, it was revealed that machine code for Intel x86 processors can also be subject to embedding [26]. Covert information can be inserted into the carrier machine code by carefully selecting functionally equivalent processor instructions. This method exploits the same principle as linguistic steganography, where the choice of words from the set of synonyms can be attributed steganographic meaning.

1.4.2 File System Steganography

The invention of a steganographic file system by Anderson, Needham, and Shamir was an eye-opener [27]. It became apparent that information can be steganographically embedded even in isolated computing environments. The main principle of steganogram preparation was similar to invisible inks—one that knew how to search could reveal the hidden encrypted files from a disk. The mechanism relies on the fact that encrypted data resemble random bits naturally present on the disk and only the ability to extract the vectors marking the file boundaries allows to locate the hidden files. Another example of a steganographic file system can be found in [28], whose authors created a steganographic file system implementation on Linux. Their invention preserves the integrity of the stored files and hides data on the disk using camouflaging with the aid of dummy hidden files and abandoned blocks.

1.4.3 Network Steganography

Currently, in the digital world, it is easily imaginable that the carrier, in which secret data are embedded, does not necessarily have to be a digital image or audio file. It may be any other file type or organizational unit of data—a packet, a frame, and so on, which naturally occurs in computer networks.

Therefore, alongside the above-mentioned types of digital steganography, currently the target of increased interest is network steganography, which is the youngest branch of information hiding. It is a fast developing field: recent years have resulted in multiple new flavors of information hiding methods, which can be exploited in various types of networks. The exploitation of protocols on the different layers of the Open Systems Interconnection Reference Model (OSI RM) is the essence of network steganography. This family of methods may utilize one or more protocols simultaneously or the relationships between them—relying on the modification of their intrinsic properties for the embedding of secret data.

The rise of network-based information hiding was thereby supported by two factors, which makes it superior to, for example, digital media steganography. First, the space for data to be transferred is not limited in the same way as it is for a given media object (e.g., a digital image file) as new data can be transmitted on demand while the capacity of a media object is limited. Second, network-level embedding of secret messages allows for leakage of information, even very slow, during long periods of time. Such transmissions are harder to analyze by digital forensic experts as only parts of the overt traffic are captured (or sometimes the traffic is not captured at all). As a result, using such ephemeral hidden data carriers makes steganographic methods harder to detect and eliminate in communication networks.

In general, the following conditions of every network steganography technique may be formulated [29]:

- $C1$: Some properties of communication protocols are modified.
- $C2$: The modification pertains to ($C2a$) properties of the protocols that cope with the intrinsic imperfectness of communication channels (errors, delays, etc.) and/or ($C2b$) properties of the protocols that define the type of information exchange (e.g., query response, file transfer, etc.) and/or adapt the form of messages (e.g., fragmentation, segmentation, etc.) to the information transmission carrier.
- $C3$: The communicating parties attempt to make the observable effects of modifications difficult to discover (e.g., make them appear as the results of the imperfectness of the communication network and/or protocols).

Conditions C1, C2, and C3 constitute a definition of network steganography techniques.

Note that if condition C1 is not fulfilled, that is, if no properties of communication protocols can be modified, then hidden communication is still possible, namely if the secret shared by the sender and receiver is of the form: messages a, b, c, . . . are interpreted as x, y, z, ... Such hidden communication cannot be discovered by observing the exchange of messages, as these are interpreted on the semantic/pragmatic level by the

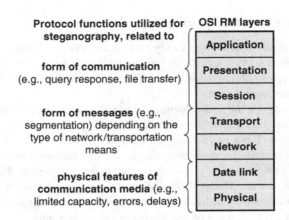

Figure 1.4. Protocol functions used for network steganography, associated with OSI RM layers. (Reproduced from [29] with permission of IEEE.)

sender and receiver. In effect, such hidden communication can be discovered only if the shared secret is disclosed. Obviously, this is not a very interesting case for research. Condition C2 refers to the fact that real-world communication protocols must realize functions (C2a) that provide the required quality-related performance of communication and functions (C2b) that govern the "logic" of communication and adapt the messages to the format of transmission carriers. If the communication functions are decomposed into functional layers, as, for example, in the OSI RM, then C2a functions are associated with lower layers and C2b functions with upper layers. In Figure 1.4, these functions, in association with OSI RM protocol layers, are characterized in a general manner. The effectiveness of a particular steganographic technique depends on how successfully C3 is fulfilled.

After a protocol or number of protocols have been chosen as a carrier for secret data, it is decided how the embedding should be performed. The first possibility is to inject the covert information into the protocol data unit (PDU)—this can be done by modifying protocol-specific fields or by inserting data into the payload, or both. Also, it is possible to modify the timing of PDUs or protocol operations. These changes may impact for example, the order of PDUs, their losses, or their relative delays. Of course, hybrid methods that utilize both modification of PDUs and timing of PDUs/operations are also possible.

The predecessor of current, more sophisticated network steganography methods was the utilization of different fields of the TCP/IP stack's protocols [30] as a hidden data carrier. The majority of the early methods concentrated on the embedding in the unused or reserved fields of protocols to convey secret data. However, nowadays, more sophisticated and advanced methods are developed, which are targeted toward specific environments or toward specific services. Recent solutions exploit

- popular Internet services, for example, Skype, BitTorrent (see Section 7.2), Google search, social media sites (see Section 7.7), and multimedia services such as IP telephony (see Section 7.1) and online, multiplayer games (see Section 7.6);

- new network protocols, such as SCTP (Stream Control Transmission Protocol) or IPv6 (see Section 7.4);
- new networking environments, for example, cloud computing or Future Internet (Internet of Things or Content-Centric Networks) (see Section 7.8);
- new networking technologies such as wireless networks, for example, wireless local area networks (WLANs) or Long-Term Evolution (LTE) (see Section 7.5);
- novel communication devices such as smartphones (see Section 7.3).

To summarize, various network services and applications can and will become the target of network steganography. The larger the proliferation of a certain service or application, the more attractive it is to be utilized as a carrier for secret data exchange. This conclusion will be further investigated in Section 1.5 where the emerging trends of information hiding will be highlighted.

1.4.4 Traffic Type Obfuscation

Traffic type obfuscation hides the type of the network traffic exchanged between multiple network entities, that is, the underlying network protocols. Early traffic type obfuscation mechanisms were developed by Internet users to bypass restrictions enforced by some ISPs on particular network protocols. For instance, some ISPs restrict the use of popular file sharing protocols such as BitTorrent to prevent sharing of pirated software and intellectual properties. As another example, some corporate networks ban the use of instant messaging services, such as Yahoo! Messenger and Skype, as well as video streaming protocols to improve their employees' productivity and protect corporate private information.

To enforce such restrictions, ISPs use different networking mechanisms [31–33] to identify (and then block) the network traffic corresponding to the disallowed protocols. Consequently, a trivial countermeasure against them is to obfuscate the type of network protocol being used, that is, to perform traffic type obfuscation.

The use of traffic type obfuscation for blocking resistance has become even more popular in the past several years. The Internet plays an ever-increasing role in the free circulation of ideas and information; events like Arab Spring in 2012 in Tunisia, Libya, and the rest of the Middle East gave strong indications that oppressive regimes can even be overthrown by the power of people mobilized to fight by organizing, communicating, and raising awareness through use of the Internet. Consequently, repressive regimes restrict their citizens' access to the Internet using different networking mechanisms, which is broadly known as Internet censorship. Recent studies [34–39] show that an effective mechanism used by the censors is to identify (and then block) the use of disallowed network protocols by analyzing network traffic at large scale. Consequently, recent censorship resistance systems [40–44] have adapted the use of traffic type obfuscation mechanisms to evade detection.

Traffic type obfuscation has alternatively been used to augment the privacy of Internet users. That is, researchers have proposed to manipulate traffic patterns in order to conceal the intent of communication. For instance, previous research [45–47] shows that an adversary can learn sensitive information about the Internet browsing activities of users (e.g., the websites they visit) by statistically analyzing their encrypted web traffic.

1.5 EMERGING TRENDS IN INFORMATION HIDING[3]

This section provides an overview about the most important trends in the two areas of information hiding in communication networks, that is, network steganography and traffic type obfuscation. A more thorough description of many methods mentioned here will be presented in Chapter 7.

1.5.1 Network Steganography

Recent years have seen the development of more and more sophisticated network steganography techniques. The current development concentrates not only on improving the stealthiness of covert channels but also on extensively enhancing the capabilities of the hidden communication [49].

As mentioned in Section 1.4.3, the predecessors of current sophisticated network steganography methods mainly focused on embedding secret data in unused or reserved fields of the TCP/IP protocol headers. However, recently we have experienced a change in the hidden data carrier selection. Now, the most favorable carrier is the one that is related to the popular Internet services, so higher layer applications and services are exploited. As a result, it is hard to spot a single steganographic communication along a vast volume of similar network traffic (''a needle in a haystack''). Moreover, steganographers tend to craft the methods in such a way that they utilize the characteristic features of these Internet services.

1.5.1.1 Popular Internet Services. Practically, any popular Internet service can be exploited by steganographers if it has enough traffic that can be altered to embed covert channels. One example is IP telephony, for which recently a number of steganographic techniques were proposed [50]. One of the steganographic methods of this kind is TranSteg (Transcoding Steganography) [51], which is based on the general idea of transcoding (lossy compression) of the voice data from a higher bit rate codec, and thus greater voice payload size, to a lower bit rate codec with smaller voice payload size (which should be performed with the least degradation in voice quality possible). The compression of the overt data creates space to insert hidden data bits while keeping the length of packets unchanged.

---

[3] The following section is written based on the publication ''Hidden and uncontrolled: on the emergence of network steganography'' by S. Wendzel, W. Mazurczyk, L. Caviglione, and M. Meier [48].

Other methods target currently popular peer-to-peer (P2P) applications such as Skype and BitTorrent. Mazurczyk et al. [52] developed a steganographic method named SkyDe (Skype Hide) that uses encrypted Skype voice packets as hidden data carrier. By taking advantage of the high correlation between speech activity and packet size, packets without voice signal can be identified and used to carry secret data by replacing the encrypted silence with secret data bits. For the BitTorrent application, a method named StegTorrent has been introduced [53]. It exploits the fact that in BitTorrent there are usually many-to-one transmissions, and that for one of its specific protocols—μTP—the header provides sequence numbers for retrieving the original sequence of packets. Thus, this characteristic feature enables to correctly order each packet at the receiver regardless of the connection it was sent from StegTorrent functions based on the intentional BitTorrent data packet resorting to achieve the desired packet sequences.

Information hiding is also possible by simply performing a series of innocent-looking Google searches. The StegSuggest steganographic method [54] targets the feature Google Suggest, which lists the 10 most popular search phrases given a string of letters a user has entered in Google's search box. The covert sender intercepts the traffic exchanged between Google servers and the covert receiver's browser. Then, when some Google search is initiated, the data traveling from Google to the covert receiver are intercepted and modified by adding a unique word to the end of each of the 10 phrases Google suggests. The choice of phrases is made from a list of 4096 common English words, so the new phrases do not look suspicious. The covert receiver extracts each added word and converts it into 10 bits of covert data using a previously shared lookup table.

1.5.1.2 New Networking Environments, Technologies, and Protocols.

The cloud computing environment can also be exploited to enable covert communication. Ristenpart et al. [55] showed that it is possible to leak confidential information between virtual machines through a wide range of channels, such as the values over time of shared-cache load, CPU load, or keystroke activity.

Covert communication channels have also been recently considered for CCNs, which are envisaged to (partially) replace current IP-based networks in the future. Ambrosin et al. [56] inspected the potential vulnerabilities that can be exploited for information hiding purposes. They focus on so-called *ephemeral covert communication*, which they define as a hidden data transfer in which the communicating parties are not exchanging any packets directly. Instead, they intentionally influence the content of the router's cache and the delay in its responses to embed and extract hidden data bits. Hidden messages are then present in the cache only for the limited time period and later they are automatically deleted from the network without any additional, required actions.

Information hiding techniques also target wireless networks. This is a very popular and dynamically evolving area of network steganography where different wireless transmission standards can be exploited for covert communication. For example, for WLANs, Szczypiorski and Mazurczyk have introduced a method called WiPad (wireless padding) [57]. The technique is based on the insertion of hidden data into the padding of frames at the physical layer of WLANs. A similar concept was also utilized, for example, for LTE [58].

Recently standardized network protocols are also prone to network steganography. For example, a detailed analysis in [59] revealed steganographic vulnerabilities in the new features and characteristics of SCTP, such as multihoming and multistreaming, that could be utilized for information hiding.

1.5.1.3 Novel Communication Devices. In the last few years, smartphones became one of the most popular communication devices. The key reasons of this huge success are

- multifunctional devices that combine many features, such as a high-resolution camera, various sensors, different communication interfaces (e.g., 3G, Bluetooth, IEEE 802.11, etc.), and Global Positioning System (GPS), into a unique and comprehensive tool;
- advanced cellular network connectivity, allowing users to interact with high-volume or delay-sensitive services while moving, for example, through the Universal Mobile Telecommunications System (UMTS), or LTE;
- large and fast growing user base igniting the development of many applications delivered from online stores, or through specialized sources on the Web.

In the context of smartphones, the possibilities for information hiding are dramatically multiplied forming a kind of a covert communication "Swiss army knife" for the following reasons: *(i)* the multimedia capabilities enable a wide variety of usable steganography carriers, such as audio, video, pictures, or quick response (QR) codes, *(ii)* the availability of a full-featured TCP/IP stack and the possibility to interact with desktop-class services allow utilizing all network steganography methods already available for standard computing devices or appliances, and *(iii)* the richness of the adopted OS permits developing sophisticated applications, such as VoIP or P2P, which can be exploited by covert channels.

It must also be noted that currently smartphones are often utilized to exchange and store high volumes of personal data, interact with online social networks (OSNs), and in the daily working practice. To handle the hardware, the typical mobile operating system (OS) has an architecture very similar to the one used on desktops; for example, Android runs a Linux kernel. However, the storing of user's sensitive data requires additional layers to ensure security and confidentiality of these data. Various techniques exist, but the most popular approach is to rely on so-called sandboxes, that is, execution environments preventing a process/application from accessing sensitive data and certain parts of the hardware.

The architectures of the current mobile OSs force steganographers to search for ways not just to exfiltrate data using network covert channels but also to acquire the data by using local covert channels in the first place. This is somewhat complicated due to the security policies enforced in mobile OSs as many manufacturers force users to install software only from verified sources; for example, iOS only permits software provided by Apple's AppStore. Therefore, network steganography methods for smartphones are typically used in combination with local covert channels [60].

More details on examples of network steganography methods for smartphones and their required cooperation with local covert channels will be provided in Chapter 7.

1.5.1.4 Control Protocols, and Adaptive and Optimization Techniques.

The tool Ping Tunnel [61] is not only capable of transferring hidden data using ICMP echo request and reply but also uses a control protocol (each block of covert data is preceded by a control protocol header). Such a control protocol usually implements reliability for the covert data transfer; that is, it enables the receiver to reorder out-of-sequence data and also enables the sender to detect and resend lost data (with the help of feedback from the receiver).

It must be noted that, in general, providing reliability for covert communication is not a trivial task due to the following reasons: *(i)* low steganographic bandwidth often makes error correction challenging; *(ii)* it is often difficult to establish a reverse channel needed for retransmissions and even if it is available care must be taken as it can impact undetectability; and *(iii)* there is lack of bit synchronization, which causes issues for error correction—this is especially important if no packet sequence numbers are provided. In this case, packet loss can cause bit deletions, which are significantly more problematic than bit errors.

Control protocols for network steganography were discussed in various academic publications in which their feature set was enhanced and will be described in the following.

A significant step was the automatic discovery of techniques that can be used between two or more peers to exchange secret data. Yarochkin et al. introduced a network environment learning (NEL) phase in which peers probe for available steganographic methods (from a set of known methods) and rule out methods relying on blocked and non-routed network protocols [62]. This also allows bypassing administrative filtering and changes in the network configuration. For instance, if a network steganographic communication is detected and an administrator blocks the utilized communication, with NEL peers can automatically switch to another steganographic method or another carrier protocol to maintain their hidden communication [49].

The next step for network steganography was to build hidden overlay networks capable of realizing a routing process. The first approach—based on the random-walk algorithm—was already presented in 2007 by Szczypiorski et al. [63]. Later, a stealth-optimized dynamic routing based on optimized link-state routing (OLSR) using size-optimized control protocols was published [64].

While control protocols for network steganography increase the capabilities of hidden communication, using them means one must optimize not only the undetectability of the transmission of the user's secret data but also the control protocol's stealthiness. This concept was initially discussed in [65], where the optimal size of a control protocol header for simultaneously used carrier protocols was proposed. Moreover, two additional protocol engineering approaches for control protocols are available today.

The first approach presented in [66] utilizes formal grammar to adjust the embedded control protocol to the utilized network protocol in which it is hidden. Therefore, the hidden control protocol must be designed to match the specification of the utilized, overt protocol to prevent anomalies due to hidden control protocol's operation. The less secret data are transferred by a steganographic method, the less attention will be raised. Therefore, the second approach for control protocol engineering is to minimize the size of the control protocol to as few bits as possible, which can be achieved by separating the components of the protocol [64].

For optimizing the communication process between peers, certain approaches exist to minimize the overhead caused by a covert communication as well as to minimize the packet count for a given amount of bytes to be transmitted over simultaneously used carriers [65].

An additional view on the improvement of the undetectability for covert communication was presented by Frączek et al. [67]. Besides classifying the techniques to improve the stealthiness of a hidden communication, the authors introduced multi-level steganography (MLS) in which at least two steganographic methods are utilized simultaneously in such a way that one method's (the upper level) network traffic serves as a carrier for the second method (the lower level)—previous approaches did only combine unnested parallel connections [65]. Such a relationship has been proven to have several potential benefits: *(i)* the lower level method can carry a cryptographic key that deciphers the steganogram carried by the upper level method; *(ii)* the lower level method can be used to provide the steganogram with integrity, and *(iii)* the lower level method may be assigned as a signaling channel for the control protocol [68].

The most recent idea is to take the distortion of the carrier that is used for transferring hidden data into account. The so-called network steganographic cost [69] measures this distortion and can be compared with measures such as the peak signal-to-noise ratio (PSNR) or mean opinion score (MOS) known from digital media steganography. By optimizing the cost of a network steganographic transmission, the stealthiness of the communication can be increased.

1.5.2 Traffic Type Obfuscation

Recent trends in traffic type obfuscation can be divided into three categories: content-based approaches, pattern-based approaches, and protocol-based approaches. Some traffic type obfuscation systems combine methods from some of the three approaches.

1.5.2.1 Content-Based Approaches. Modern state-of-the-art network firewalls are equipped with deep-packet inspection (DPI) technologies, which allows them to deeply scan network traffic in various OSI layers. Many open-source DPI systems (e.g., Bro [70], Yaf [71], and nProbe [72]) use regular expressions (regexes) to classify network protocols. DPIs look for content signatures that identify a particular network protocol; hence, they can be used to identify and block specific network protocols, for example, Skype and Tor protocols.

Content-based traffic type obfuscation schemes aim at defeating DPI technologies by manipulating packet contents so as to remove content signatures that reveal the use of a particular network protocol. For instance, Tor [73] uses various content-based obfuscation mechanisms to hide the use of its protocol. Example content-based obfuscators for Tor include Obfsproxy [74], SkypeMorph [41], StegoTorus [42], and FTE [75].

1.5.2.2 *Pattern-Based Approaches.*

Past work [45,76–79] demonstrates the feasibility of protocol classification by analyzing traffic patterns, for example, analysis of patterns of packet timings and packet sizes. Pattern-based traffic obfuscation schemes aim at concealing the type of traffic by perturbing patterns of network traffic. For instance, Wright et al. [79] propose two approaches that aim at mimicking another network protocol based on prerecorded network traces. They proposed the two schemes of direct target sampling and traffic morphing, which modify packet sizes of a network flow in order to obfuscate its type.

1.5.2.3 *Protocol-Based Schemes.*

The third trending class of protocol type obfuscation includes protocol-based schemes. These techniques modify the protocol behavior such as the handshaking mechanism and subprotocol dependencies, in order to obfuscate protocol types. For instance, CensorSpoofer [40] is a system for blocking resistance web browsing that uses protocol-based obfuscation. It obfuscates the web protocol to look like a P2P protocol such as Voice over IP (VoIP). FreeWave [43] is another protocol-based obfuscation scheme, which makes a network protocol look like Skype. Other example systems include SkypeMorph [41], StegoTorus [42], and CloudTransport [44].

1.6 APPLICATIONS OF INFORMATION HIDING AND RECENT USE CASES

In general, information hiding techniques can be treated as a double-edged sword depending on who uses them and how. Information hiding can have benefits; for example, it can help maintain privacy, but it can also pose a threat to individuals, societies, and states when used for malicious purposes. The trade-off between the benefits and threats involves many complex ethical, legal, and technological issues that require consideration in a broader context, which is beyond the scope of this book.

The use cases for information hiding in general are manifold but as indicated in recent research are mainly linked to illicit purposes [6]. Before discussing malicious cases that attracted most attention in worldwide media, some interesting examples of benign applications are presented first.

1.6.1 Benign Applications of Information Hiding

In 2010, Ibaida and Khalil [80] proposed to embed patient information into electrocardiogram (ECG) signals so that it can only be accessed by healthcare workers who have the correct credentials. The method can also be applied to other medical monitoring devices. It utilizes a wavelet-based steganography together with encryption and scrambling techniques to protect patient's sensitive data.

Palacios et al. [81] propose to use living organisms as carriers of secret data using a kind of a biotech invisible ink. The main idea is to write and encode data using arrays of genetically engineered strains of *Escherichia coli* with fluorescent proteins as phenotypic markers. Generated in this way, messages consist of a matrix of spots generated by seven strains of *E. coli*, each having a different color. The coding scheme for these arrays relies on strings of paired digits that represent an alphanumeric character. Among the main potential applications are forgery-resistant bacterial barcodes and watermarks.

Another example (mentioned earlier) is the utilization of steganography for printer identification purposes [82]. Xerox DocuColor color laser printers mark each printout with the date, time, and printer serial number in a form of forensic tracking codes. It is achieved by printing a rectangular grid of 15×8 miniscule yellow dots on every color page printed by a user. These dots allow encoding up to 100 bits of tracking information. Because of their limited contrast with the background, the forensic dots are invisible to the naked eye under white light—they appear only when magnified or illuminated by blue instead of white light.

Legitimate uses also include circumvention of web censorship and surveillance by oppressive regimes, enabling computer/network forensic methods and copyright protection. For IP telephony information hiding techniques can be used to improve its resistance to packet losses and improve voice quality [83,84], extend communication bandwidth [85], or provide means for secure cryptographic key distribution [86].

Moskowitz et al. showed that imperfections in anonymous communications are effectively covert side channels that can be utilized to break the anonymization [87] (however it must be noted that breaking anonymization is not always a benign application of information hiding and it depends mainly on the context). Furthermore, covert channels have been used for breaking anonymization in several scenarios. For example, Xu et al. proposed an attack on traffic trace file anonymization using covert channels [88] and Murdoch used covert channels to reveal the identity of hidden servers inside anonymization networks [89]. Covert channels can also be used for user authentication.

So-called "port knocking" techniques use covert channels for sending authentication data in order to open firewall ports for authorized users [90].

1.6.2 Malicious Applications of Information Hiding

As mentioned earlier, in communication networks the main application of information hiding is clandestine communication and the purposes for establishing hidden data exchange often fall into the category of legal or illicit activities. The illegal purposes of steganography range from criminal communication, confidential data exfiltration from guarded systems, cyber weapon exchange and control, to industrial espionage.

One of the most infamous examples is the alleged use of steganographic methods by terrorists while planning the attack carried out in the United States on the 11th of September, 2001 [91,92]. Press articles described fake eBay listings and other publicly available websites in which routinely altered pictures contained hidden data, which may have been instructions regarding the plot. It seems that such communication could have passed unnoticed for as long as 3 years [93]. The link to 9/11 was never proved or disproved, but after those reports, the interest in steganographic techniques and their detection greatly increased.

In 2002, "Operation Twins" culminated in the capture of criminals associated with the "Shadowz Brotherhood" group, a worldwide Internet pedophile organization responsible for distribution of child pornography [94]. The group utilized encryption but also digital image steganography to hide pornographic images within innocent-looking images.

A large fraction of the security breaches in 2011 was attributed to the "Operation Shady RAT" [95] that began in 2006. It included a wave of attacks against numerous institutions (including defense contractors, the United Nations, and the International Olympic Committee). By means of phishing emails, a victim was tricked into installing a back-door, which received commands and executable code hidden in innocent-looking HTML or JPEG files [95]. In numerous cases, the side channel to the confidential resources remained accessible for months, thus making this a severe security breach. The villains were so daring that they did not even put much effort into obscuring the fact that information hiding techniques were involved in the attack. One of the pictures used as a vector for control commands was the famous "Lena," a cropped picture of a Playboy model, which is the standard test image for any digital image processing or steganographic algorithm.

According to a rumor, in 2008, someone at the U.S. Department of Justice smuggled sensitive financial data out of the agency by embedding the data in several image files [96].

In 2010, the Russian spy ring of the so-called "illegals" was discovered by the FBI in America. It is believed that the spy network started using digital image steganography in as early as 2005 in order to leak classified information from the United States to Moscow by posting modified digital images with secret messages to public websites.[4]

A suspected member of Al-Qaeda was arrested in Berlin in May 2011 with a memory card containing a password-protected folder. Computer forensic experts from the German Federal Criminal Police (BKA) were able to reveal its content, which turned out to be pornographic videos. But there was more—within these files, forensic investigators discovered more than 100 documents detailing the terrorists' current operations and future plans [97].

The mushrooming incidents involving the use of information hiding triggered an official recognition of the problem. In the 2006 Federal Report [98], steganography has been named as one of the major threats of the present-day networks, whose significance is predicted to increase. One of the solutions to alleviate the risks connected with this technique is to become acquainted with the evolution of steganography and, consequentially, predict its further development. This need had been recognized by the academic world in the early 1980s already when steganography started to gain popularity.

As already mentioned, information hiding techniques can be utilized by malware to cloak its existence, thereby making its detection harder. This trend has been enforced

---

[4] The U.S. Justice Department complaint is available at
http://www.justice.gov/sites/default/files/opa/legacy/2010/06/28/062810complaint2.pdf.

by the advancements in security systems over the last 15 years that have forced malware developers to investigate new possibilities to make their "products" stealthier.

Mazurczyk and Caviglione [99] discovered that nearly all of the information hiding-capable malware has been discovered between 2011 and 2015. Existing hiding-capable malware can be classified according to the methodology used to implement the covert communication. They identified three major groups:

- Methods hiding information by modulating the status of shared hardware/software resources.
- Methods injecting secret data into network traffic.
- Methods embedding secret data by modifying the structure of a digital file or by using digital media steganography, for example, by manipulating pixels of a picture or samples of an audio file.

The second and third groups of methods contain techniques that are primarily used to increase the stealthiness of communications carrying commands or leaked data that are currently mainly observed in malware targeting desktops. The first group of methods instead embraces mechanisms to bypass a security perimeter, such as a sandbox, or to enable communications from/to an isolated source/destination (e.g., two disconnected devices located on the same workbench). In this case, the prime targets are smartphones and mobile devices. In the following subsections, for each group, the three most meaningful examples are described.

1.6.2.1 Methods Hiding Information by Modulating the Status of Shared Hardware/Software Resources. The increasing attention of researchers combined with the open-source nature of Android has allowed the development of many examples of proof-of-concept information hiding-capable mobile malware. A prime example is Soundcomber [100], which covertly transmits the keys pressed during a call (e.g., when entering a PIN for a bank service). Notably, it uses information hiding to bypass the security framework of mobile OSs. In fact, the malware could have insufficient privileges to access the network to exfiltrate data. Thus, it can use a "colluding" application to leak data from the device. Soundcomber utilizes several information hiding methods to create local covert channels between different processes on the device. The proposed techniques exploit the most popular functionalities of the smartphone such as vibration or volume settings (one process is differencing the status of the vibration/volume, and the other infers secret data bits from this event), screen state (secret bits are transferred by acquiring and releasing the wake-lock permission that controls the screen state), and file locks (secret data are exchanged between the processes by competing for a file lock).

As hinted, another relevant field in which information hiding can be used is the covered transmission of data from/to devices physically isolated from other peers. For instance, the mechanism of Deshotels [101] uses the standard speaker of a smartphone to transmit data via ultrasonic sounds. This technique can cover distances up to 30 m with a rate of 9 bit/s. Similarly, AirHopper [102] enables infected devices to communicate by modulating the load of the graphics processing unit (GPU) to emit electromagnetic signals. In this case, the coverage is reduced to 7 m, but the rate is in the range of 100–500 bit/s.

Finally, Hasan et al. [103] demonstrate a method to trigger attacks on a large population of infected smartphones located in the same geographical area. Latent malware could be activated by using built-in sensors listening to ad-hoc hidden stimuli, such as a song with a particular pattern, vibrations from a subwoofer, or the ambient light from a TV or a monitor.

1.6.2.2 Methods Injecting Secret Data into Network Traffic.

In 2011, Symantec announced the discovery of the worm W32.Morto [104], which propagates using a vulnerability of the Remote Desktop Protocol. To communicate with the command and control (C&C) servers, that is, nodes used for coordination and information gathering purposes, it used records of the Domain Name System (DNS). Specifically, W32.Morto exploits the DNS TXT record, introduced originally to contain human-readable text. W32.Morto queries for a TXT record (not for a domain to IP lookup), and the returned data are validated and decrypted. The obtained information typically yields a binary signature and an IP address from which the worm can retrieve malware code to execute.

More recently, Linux.Fokirtor was identified, a Trojan that opens a backdoor and allows an attacker to remotely compromise a host. It was reported [105] that the malware was utilized in May 2013 to attack one of the large hosting providers and was focused on stealing confidential customer information such as credentials and emails. As cybercriminals realized that their target network is generally well protected, to avoid setting off any alarm bells, they hid malware communications in innocent Secure Shell (SSH) and other server process network traffic. In addition to this information hiding technique, Linux.Fokirtor utilized the Blowfish encryption algorithm to encrypt the communications with its C&C server.

In November 2014, Regin [106] has taken malware stealthiness a step further. It utilized many sophisticated mechanisms: anti-forensic capabilities, a custom-built encrypted virtual file system, and alternative encryption (RC5 variant). It also exploits information hiding in network traffic to covertly communicate with its C&C server by tunneling secrets in ICMP traffic and embedding commands in HTTP cookies or in custom TCP segments and UDP datagrams.

1.6.2.3 Methods Embedding Secret Data by Modifying the Structure of a Digital File or by using Digital Media Steganography.

In the second half of 2011, the Hungarian Laboratory of Cryptography and System Security discovered a malware generating strange files with the prefix ''~DQ''; hence, it has been named Duqu [107]. It bears many resemblances to the famous Stuxnet [108], which was probably developed to attack Iran's nuclear infrastructure. It is essentially considered the precursor to a future Stuxnet-like attack. Duqu's main aim was to gather information about industrial control systems. To exfiltrate secrets, Duqu encrypted the data, appended it at the end of innocent digital images, and then sent it over the Internet to a C&C server. This approach postponed its detection because the images containing leaked information were hidden within the bulk of licit digital pictures. In the same period, a variant of Alureon used a comparable technique.

In February 2014, a malware called Lurk [109] was discovered spreading via websites using <iframes> or an exploit in Adobe Flash. A thorough analysis revealed the use of steganography to embed encrypted URLs within an image by manipulating pixels. Such information is then used to retrieve an additional payload.

Another approach has been utilized in a variant of the Trojan.Zbot [110], which was first detected in 2014. This version was able to download innocent-looking JPEG images (specifically, depicting a sunset or some cats) containing a list of IP addresses to be inspected, mainly pointing at financial institutions. Once the user visits any of the listed destinations, the malware proceeds to steal his/her confidential information, such as access credentials.

Regin, Duqu, and Lurk are some real-life examples of what security experts should expect as the daily routine of the future. In fact, even if the information hiding methods utilized by malware found in the wild are not yet very sophisticated, they could become dangerous in the next few years if they advance to the state-of-the-art academic hiding methods.

1.7 COUNTERMEASURES FOR INFORMATION HIDING TECHNIQUES

As mentioned in Section 1.1, there is an analogy between the ongoing evolution of both offensive and defensive techniques in nature and in the area of information hiding. In both cases, we are witnessing an *arms race*. The general rule is that for every information hiding technique developed, sooner or later, a defensive scheme appears. Then, in response to the developed countermeasure, the steganographic technique is enhanced, but usually only to the point where it is stealthy again given the existing countermeasure.

In general, countermeasures for information hiding methods include ways to [111]:

- detect covert communication;
- prevent hidden data exchange;
- limit the effectiveness of information hiding techniques, if they cannot be eliminated completely.

A major problem for developing effective countermeasures is the great diversity of existing information hiding techniques. In communication networks, secret data can be embedded not only by modifying various network protocols or various types of digitally transmitted media content (such as audio streams, video streams, or HTML content), but also by using a large number of hiding techniques. In other words, not only are there a lot of potential, suitable carriers available, but there are also multiple ways to utilize each carrier for hidden data exchange. Even if methods that modify digital payload are excluded, there exist more than one hundred techniques for secret data transfer using network protocols, such as protocol header elements or the timing of network packets [48].

Due to the increasing complexity and huge diversity of information hiding techniques, it is extremely difficult to develop accurate and effective countermeasures that can be broadly applied, that is, countermeasures that work for many different covert channels. Hence, the existing countermeasures are usually highly specialized and only work for one or a few hiding techniques. A more general approach for countering clandestine communication—one that will not deal with each information hiding method alone—is needed. A recently proposed approach, which is still in its infancy, is to classify hiding techniques into so-called patterns (abstract descriptions) and develop countermeasures for these patterns [112]. With this approach, one countermeasure can be used for a large number of hiding techniques represented by one particular pattern.

The most common approach today is to use a combination of machine learning methods to detect covert channels and traffic normalizers to disrupt covert transmissions. Some countermeasures are already integrated in data leakage protection (DLP) products available on the market; however, the effectiveness of these solutions is often unclear. Given that information hiding techniques are becoming increasingly sophisticated and complex, the detection, limitation, and prevention is becoming even more challenging. New research must come up with fundamental approaches to counter the most recent and future techniques. Afterward, products must be developed to enable protection in operational network environments.

The existing countermeasures for information hiding techniques will be presented in more detail in Chapter 8.

1.8 POTENTIAL FUTURE TRENDS IN INFORMATION HIDING

Nils Bohr once amusingly said that "prediction is very difficult, especially if it's about the future." As forecasting the future of information hiding is a difficult thing to do, below we present some probable directions in which such techniques for communication networks *might* evolve.

First, it can be expected that the application of the information hiding techniques will lead to even more sophisticated malware that will be stealthier and therefore harder to detect than nowadays (the first symptoms have already been observed as described in Section 1.6) forming a new type of advanced persistent threat (APT). Deploying dynamic overlay routing and adaptive network covert channel techniques, malware will moreover be harder to prevent on the network level. A host-based detection of malware is thus important when a network-level detection and prevention is not feasible in these situations. Today's C&C channels in botnets already possess a comprehensive feature set and adapting these features to the context of information hiding methods is only one step in the malware evolution.

Second, an increasing stealthiness of malware communications on smartphones can be observed. While there is not a major effort in developing novel network covert channels especially crafted for smartphones, recent trends take advantage of the device offloading features, especially those using the cloud. In fact, to bypass storage or battery limitation of devices, some operations are delegated to a remote server farm. In this perspective, new applications producing traffic potentially exploitable for network steganography are becoming available. We mention, among others, voice-based services such as Google Now and Siri or cloud storage platforms such as Google Drive and Dropbox. Even if many frameworks use protocols/techniques already exploited in network steganography (e.g., HTTP), the huge traffic volumes and the degree of sophistication of many services will represent a great challenge to detect or eliminate the covert communication.

Third, a spread of information hiding techniques to the new domains, especially when combined with existing malware, can be envisioned. One example in this regard is the potential to form novel botnets consisting of smart buildings instead of computers (so-called smart building botnets), allowing the remote mass surveillance and remote control of smart cities [113]. Information hiding methods can increase the stealthiness of mass surveillance in such situations, especially when the number of bots in a smart building botnet is high.

Fourth, a utilization of information hiding could increase the stealthiness of illegal data exchange, including communications already found within darknets today, such as the exchange of child pornographic material or hitman hiring.

Fifth, it can be expected that information hiding techniques will significantly influence industrial espionage. While nowadays the exfiltration of sensitive data is based on simple techniques in many cases—for example, data is exfiltrated via email or USB stick—recent steganography use cases (for details see Section 1.6) have shown that digital image steganography has been indeed applied to exfiltrate data out of organizational environments. Moreover, with information hiding techniques confidential data exfiltration can be realized not only in a stealthy but also in a continuous manner, for example, by intentionally leaking a small amount of data per hour.

Sixth, traffic type obfuscation mechanisms may evolve by going higher in the network protocol stack. Early traffic type obfuscation systems, such as Traffic Morphing [79] and StegoTorus [42], work at the network layer as they try to mimic a target network protocol. Recent studies [34,35] show that modern, powerful network monitoring equipment of tomorrow will be able to counter these systems, suggesting to obfuscate network traffic in the higher layers of the network protocol stack, for example, in the application layer. FreeWave [43] is an example for such systems.

1.9 SUMMARY

Information hiding techniques have their roots in nature and they have been utilized by humankind for ages. The methods have been evolving throughout the ages but the aims remained the same: to hide some secret information from untrusted parties or to enable covert communication. The latter grew in importance with the introduction of communication networks where many new possibilities of data hiding emerged.

Information hiding can be utilized for both benign and malicious purposes; currently, the rising trend is to equip malware with covert communication capabilities for increased stealthiness. The complexity and richness of continuously emerging new services and protocols guarantee that there still will be a lot of new opportunities to hide secret data. This also urgently demands for more universal countermeasures that will be able to detect not only a single method (as it is today) but a larger number of information hiding techniques.

1.10 ORGANIZATION OF THE BOOK

The book is divided into eight chapters that cover the most important aspects of information hiding techniques for communication networks.

Chapter 1 is written mostly in a tutorial style so that even a general reader will be able to easily grasp the basic concepts of information hiding, their evolution throughout the history, and their importance especially when utilized in networking environments. It also contains many examples of applications of modern information hiding for criminal and legitimate purposes, and it highlights current development trends and potential future directions.

Chapter 2 discusses the existing terminology and its evolution in the information hiding field. It introduces a new classification of data hiding techniques; however, our new classification builds on existing concepts. The chapter then introduces the two main subfields: network steganography and traffic type obfuscation methods. The chapter concludes with a description of the model for hidden communication and related communication scenarios. It also highlights potential countermeasures.

Chapter 3 describes in detail different flavors of network steganography. Three main types of techniques are distinguished and then characterized: hiding information in protocol modifications, in the timing of network protocols, and hybrid methods.

Chapter 4 introduces techniques that improve the resiliency and undetectability of network steganography methods. These techniques are usually implemented by so-called control protocols. The chapter discusses their features, highlights the design of known control protocols, and discusses control protocol-specific engineering methods.

Chapter 5 concentrates on traffic type obfuscation techniques that allow to hide the type of the network traffic exchanged between two (or multiple) network entities, that is, the underlying network protocol. Typical applications of these methods are two-fold: blocking resistance or privacy protection. The chapter presents a classification of traffic type obfuscation techniques and covers the most important of these techniques in detail.

Chapter 6 focuses on network flow watermarking. Network flow watermarking manipulates the traffic patterns of a network flow, for example, the packet timings, or packet sizes, in order to inject an artificial signal into that network flow—a watermark. This watermark is primarily used for linking network flows in application scenarios where packet contents are striped of all linking information.

Chapter 7 presents most recent examples and applications of information hiding in communication networks with a focus on current covert communication methods for popular Internet services. This includes hiding information in virtual worlds (e.g., multiplayer online games), IP telephony, wireless networks and modern mobile devices, and P2P networks and their global services such as BitTorrent and Skype. Additionally, we discuss potential steganographic methods for social networks and the Internet of Things (e.g., building automation systems).

Chapter 8 discusses potential countermeasures against network steganography. The chapter describes different types of techniques that lead to the detection, prevention, and limitation of hidden communication.

Chapter 9 concludes the book.

REFERENCES

1. W. Mazurczyk and E. Rzeszutko. Security—a perpetual war: lessons from nature. *IT Professional*, 17(1):16–22, 2015.

2. G. D. Ruxton, T. N. Sherratt, and M. P. Speed. *Avoiding Attack: The Evolutionary Ecology of Crypsis, Warning Signals and Mimicry*. Oxford University Press, 2004.

3. R. I. Vane-Wright. A unified classification of mimetic resemblances. *Biological Journal of the Linnean Society*, 8(1):25–56, 1976.

4. M. A. Ramsier, A. J. Cunningham, G. L. Moritz, J. J. Finneran, C. V. Williams, P. S. Ong, S. L. Gursky-Doyen, and N. J. Dominy. Primate communication in the pure ultrasound. *Biology Letters*, 8(4):508–511, 2012.

5. T. H. Chiou, L. M. Mathger, R. T. Hanlon, and T. W. Cronin. Spectral and spatial properties of polarized light reflections from the arms of squid (*Loligo pealeii*) and cuttlefish (*Sepia officinalis* L.). *The Journal of Experimental Biology*, 210:3624–3635, 2007.

6. E. Zielińska, W. Mazurczyk, and K. Szczypiorski. Trends in steganography. *Communications of the ACM*, 57(3):86–95, 2014.

7. A. D. Selincourt. *Herodotus: The Histories*. Penguin Books, 1954.

8. S. Singh. *The Code Book: The Secret History of Codes and Codebreaking*. Fourth Estate, 2000.

9. D. Smith. Number games and number rhymes: the great number game of dice. *The Teachers College Record*, 13(5):39–53, 1912.

10. B. Rudin and R. Tanner. *Making Paper: A Look into the History of an Ancient Craft*. Rudins, 1990.

11. F. A. Petitcolas, R. Anderson, and M. Kuhn. Information hiding—a survey. *Proceedings of the IEEE*, 87(7):1062–1078, 1999.

12. I. Cox. *Digital Watermarking and Steganography*. Morgan Kaufmann, 2008.

13. R. Bergmair. A comprehensive bibliography of linguistic steganography. In *Proceedings of the SPIE International Conference on Security, Steganography, and Watermarking of Multimedia Contents*, 2007.

14. W. White. *The Microdot: History and Application*. Phillips Publications, 1992.

15. D. Kahn. The history of steganography. In R. Anderson, editor, *Information Hiding*, Vol. 1174 of *Lecture Notes in Computer Science*, pp. 1–5. Springer, Berlin, 1996.

16. H. Markey and G. Antheil. Secret communication system. U.S. Patent 2,292,387, 1942.

17. G. J. Simmons. The prisoners' problem and the subliminal channel. In D. Chaum, editor, *Advances in Cryptology*, pp. 51–67. Springer, New York, 1984.

18. N. F. Johnson and S. Jajodia. Steganalysis of images created using current steganography software. In *Information Hiding*, Vol. 1525 of *Lecture Notes in Computer Science*, pp. 273–289. Springer, Berlin, 1998.

19. W. Bender, D. Gruhl, N. Morimoto, and A. Lu. Techniques for data hiding. *IBM Systems Journal*, 35(34):313–336, 1996.

20. A. Cheddad, J. Condell, K. Curran, and P. M. Kevitt. Digital image steganography: survey and analysis of current methods. *Signal Processing*, 90(3):727–752, 2010.

21. Y. Wang and E. Izquierdo. High-capacity data hiding in MPEG-2 compressed video. In *Proceeding of the 9th International Workshop on Systems, Signals and Image Processing*, pp. 212–218, 2002.

22. C. Xu, X. Ping, and T. Zhang. Steganography in compressed video stream. In *1st International Conference on Innovative Computing, Information and Control (ICICIC 06)*, Vol. 1, pp. 269–272, August 2006.

23. R. Anderson. Stretching the limits of steganography. In R. Anderson, editor, *Information Hiding*, Vol. 1174 of *Lecture Notes in Computer Science*, pp. 39–48. Springer, Berlin, 1996.

24. A. Castiglione, A. D. Santis, U. Fiore, and F. Palmieri. An asynchronous covert channel using spam. *Computers & Mathematics with Applications*, 63(2):437–447, 2012.

25. K. Bennett. Linguistic steganography: survey, analysis, and robustness concerns for hiding information in text. Technical Report, CERIAS Tech Report, Purdue University, 2004.

26. R. El-Khalil and A. Keromytis. Hydan: hiding information in program binaries. In J. Lopez, S. Qing, and E. Okamoto, editors, *Information and Communications Security*, Vol. 3269 of *Lecture Notes in Computer Science*, pp. 187–199. Springer, Berlin, 2004.

27. R. Anderson, R. Needham, and A. Shamir. The steganographic file system. In *Information Hiding*, Vol. 1525 of *Lecture Notes in Computer Science*, pp. 73–82. Springer, Berlin, 1998.

28. H. Pang, K.-L. Tan, and X. Zhou. StegFS: a steganographic file system. In *Proceedings of the 19th International Conference on Data Engineering*, pp. 657–667, March 2003.

29. J. Lubacz, W. Mazurczyk, and K. Szczypiorski. Principles and overview of network steganography. *IEEE, Communications Magazine*, 52(5):225–229, 2014.

30. C. Rowland. Covert channels in the TCP/IP protocol suite. *First Monday*, 2(5), 1997.

31. Y.-D. Lin, C.-N. Lu, Y.-C. Lai, W.-H. Peng, and P.-C. Lin. Application classification using packet size distribution and port association. *Journal of Network and Computer Applications*, 32(5):1023–1030, 2009.

32. D. Bonfiglio, M. Mellia, M. Meo, D. Rossi, and P. Tofanelli. Revealing Skype traffic: when randomness plays with you. *ACM SIGCOMM Computer Communication Review*, 37(4):37–48, 2007.

33. K. Bauer, D. McCoy, D. Grunwald, and D. Sicker. BitStalker: accurately and efficiently monitoring BitTorrent traffic. In *1st IEEE International Workshop on Information Forensics and Security (WIFS 2009)*, pp. 181–185. IEEE, 2009.

34. A. Houmansadr, C. Brubaker, and V. Shmatikov. The Parrot is Dead: observing unobservable network communications. In *34th IEEE Symposium on Security and Privacy*, Oakland, CA, 2013.

35. J. Geddes, M. Schuchard, and N. Hopper. Cover your ACKs: pitfalls of covert channel censorship circumvention. In *Proceedings of the 2013 ACM SIGSAC Conference on Computer and Communications Security (CCS)*, pp. 361–372. ACM, 2013.

36. P. Winter and S. Lindskog. How the Great Firewall of China is blocking Tor. In *FOCI*, 2012.

37. T. Wilde. Knock knock knockin' on bridges' doors. https://blog.torproject.org/blog/knock-knock-knockin-bridges-doors, 2012.

38. Ten ways to discover tor bridges. https://blog.torproject.org/blog/research-problems-ten-ways-discover-tor-bridges.

39. D. Adami, C. Callegari, S. Giordano, M. Pagano, and T. Pepe. Skype-Hunter: a real-time system for the detection and classification of Skype traffic. *International Journal of Communication Systems*, 25(3):386–403, 2012.

40. Q. Wang, X. Gong, G. Nguyen, A. Houmansadr, and N. Borisov. CensorSpoofer: asymmetric communication using IP spoofing for censorship-resistant web browsing. In *Proceedings of the ACM Conference on Computer and Communications Security (CCS)*, 2012.

41. H. Moghaddam, B. Li, M. Derakhshani, and I. Goldberg. SkypeMorph: protocol obfuscation for Tor bridges. In *Proceedings of the ACM Conference on Computer and Communications Security (CCS)*, 2012.

42. Z. Weinberg, J. Wang, V. Yegneswaran, L. Briesemeister, S. Cheung, F. Wang, and D. Boneh. StegoTorus: a camouflage proxy for the Tor anonymity system. In *Proceedings of the ACM Conference on Computer and Communications Security (CCS)*, 2012.

43. A. Houmansadr, T. Riedl, N. Borisov, and A. Singer. I want my voice to be heard: IP over Voice-over-IP for unobservable censorship circumvention. In *Network and Distributed System Security Symposium (NDSS)*, 2013.

44. C. Brubaker, A. Houmansadr, and V. Shmatikov. CloudTransport: using cloud storage for censorship-resistant networking. In *Privacy Enhancing Technologies Symposium (PETS)*, 2014.

45. K. P. Dyer, S. E. Coull, T. Ristenpart, and T. Shrimpton. Peek-a-boo, I still see you: why efficient traffic analysis countermeasures fail. In *2012 IEEE Symposium on Security and Privacy (SP)*, pp. 332–346. IEEE, 2012.

46. D. Herrmann, R. Wendolsky, and H. Federrath. Website fingerprinting: attacking popular privacy enhancing technologies with the multinomial naïve-Bayes classifier. In *Proceedings of the 2009 ACM Workshop on Cloud Computing Security*, pp. 31–42. ACM, 2009.

47. M. Liberatore and B. N. Levine. Inferring the source of encrypted http connections. In *Proceedings of the 13th ACM Conference on Computer and Communications Security*, pp. 255–263. ACM, 2006.

48. S. Wendzel, W. Mazurczyk, L. Caviglione, and M. Meier. Hidden and uncontrolled: on the emergence of network steganographic threats. In H. Reimer, N. Pohlmann, and W. Schneider, editors, *ISSE 2014 Securing Electronic Business Processes*, pp. 123–133. Springer Fachmedien Wiesbaden, 2014.

49. S. Wendzel and J. Keller. Hidden and under control. *Annals of telecommunications (Annales des Telecommunications)*, 69(7–8):417–430, 2014.

50. W. Mazurczyk. VoIP steganography and its detection—a survey. *ACM Computing Surveys*, 46(2):20:1–20:21, 2013.

51. W. Mazurczyk, P. Szaga, and K. Szczypiorski. Using transcoding for hidden communication in IP telephony. *Multimedia Tools and Applications*, 70(3):2139–2165, arxiv.org/pdf/1406.2519, 2014.

52. W. Mazurczyk, M. Karaś, and K. Szczypiorski. SkyDe: a Skype-based steganographic method. *International Journal of Computers, Communications & Control (IJCCC)*, 8(3):389–400, 2013.

53. P. Kopiczko, W. Mazurczyk, and K. Szczypiorski. StegTorrent: a steganographic method for the P2P file sharing service. In *Proceedings of the 2013 IEEE Security and Privacy Workshops (SPW '13)*, pp. 151–157. IEEE Computer Society, Washington, DC, 2013.

54. P. Białczak, W. Mazurczyk, and K. Szczypiorski. Sending hidden data via Google Suggest. In *Proceedings of the International Conference on Telecommunication Systems, Modeling and Analysis (ICTSM 2011)*, pp. 121–131, 2011.

55. T. Ristenpart, E. Tromer, H. Shacham, and S. Savage. Hey, you, get off of my cloud: exploring information leakage in third-party compute clouds. In *Proceedings of the 16th ACM Conference on Computer and Communications Security (CCS '09)*, pp. 199–212. ACM, New York, 2009.

56. M. Ambrosin, M. Conti, P. Gasti, and G. Tsudik. Covert ephemeral communication in named data networking. In *Proceedings of the 9th ACM Symposium on Information, Computer and Communications Security, ASIA CCS '14*, pp. 15–26. ACM, New York, 2014.

57. K. Szczypiorski and W. Mazurczyk. Steganography in IEEE 802.11 OFDM symbols. *Security and Communication Networks*, doi: 10.1002/sec.306, 2011.

58. I. Grabska and K. Szczypiorski. Steganography in long term evolution systems. In *2014 IEEE Security and Privacy Workshops (SPW)*, pp. 92–99, May 2014.

59. W. Fraczek, W. Mazurczyk, and K. Szczypiorski. Hiding information in a stream control transmission protocol. *Computer Communications*, 35(2):159–169, 2012.

60. W. Mazurczyk and L. Caviglione. Steganography in modern smartphones and mitigation techniques. *IEEE Communications Surveys & Tutorials*, PP(99):1–1, 2014.

61. D. Stødle. Ping Tunnel—for those times when everything else is blocked. Bolg, 2005.

62. F. Yarochkin, S.-Y. Dai, C.-H. Lin, Y. Huang, and S.-Y. Kuo. Towards adaptive covert communication system. In *14th IEEE Pacific Rim International Symposium on Dependable Computing (PRDC 08)*, pp. 153–159, December 2008.

63. K. Szczypiorski, I. Margasiński, and W. Mazurczyk. Steganographic routing in multi agent system environment. *Journal of Information Assurance and Security*, 2:235–243, 2007.

64. P. Backs, S. Wendzel, and J. Keller. Dynamic routing in covert channel overlays based on control protocols. In *2012 International Conference for Internet Technology And Secured Transactions*, pp. 32–39, December 2012.

65. S. Wendzel and J. Keller. Low-attention forwarding for mobile network covert channels. In B. De Decker, J. Lapon, V. Naessens, and A. Uhl, editors, *Communications and Multimedia Security*, Vol. 7025 of *Lecture Notes in Computer Science*, pp. 122–133. Springer, Berlin, 2011.

66. S. Wendzel and J. Keller. Systematic engineering of control protocols for covert channels. In B. De Decker and D. Chadwick, editors, *Communications and Multimedia Security*, Vol. 7394 of *Lecture Notes in Computer Science*, pp. 131–144. Springer, Berlin, 2012.

67. W. Fraczek, W. Mazurczyk, and K. Szczypiorski. How hidden can be even more hidden? In *3rd International Conference on Multimedia Information Networking and Security (MINES)*, pp. 581–585, November 2011.

68. W. Fraczek, W. Mazurczyk, and K. Szczypiorski. Multilevel steganography: improving hidden communication in networks. *Journal of Universal Computer*

Science, 18(14):1967–1986, 2012. http://www.jucs.org/jucs_18_14/multilevel_steganography_improving_hidden.

69. W. Mazurczyk, S. Wendzel, I. A. Villares, and K. Szczypiorski. On importance of steganographic cost for network steganography. *International Journal of Security and Communication Networks*, 2014.

70. V. Paxson. Bro: a system for detecting network intruders in real-time. *Computer Networks*, 31(23):2435–2463, 1999.

71. C. Inacio and B. Trammell. Yaf: yet another flowmeter. In *Proceedings of the Large Installation System Administration Conference (LISA)*, 2010.

72. L. Deri .nProbe: an open source netflow probe for gigabit networks. In *Proceedings of TERENA Networking Conference*, 2003.

73. R. Dingledine, N. Mathewson, and P. Syverson. Tor: The second-generation onion router. In *USENIX Security*, 2004.

74. A simple obfuscating proxy. https://www.torproject.org/projects/obfsproxy.html.en.

75. D. Fifield, N. Hardison, J. Ellithrope, E. Stark, R. Dingledine, D. Boneh, and P. Porras. Evading censorship with browser-based proxies. In *Privacy Enhancing Technologies Symposium (PETS)*, 2012.

76. X. Cai, X. C. Zhang, B. Joshi, and R. Johnson. Touching from a distance: website fingerprinting attacks and defenses. In *Proceedings of the 2012 ACM Conference on Computer and Communications Security*, pp. 605–616. ACM, 2012.

77. X. Cai, R. Nithyanand, T. Wang, R. Johnson, and I. Goldberg. A systematic approach to developing and evaluating website fingerprinting defenses. In *ACM Conference on Computer and Communications Security (CCS)*, 2014.

78. M. Juarez, S. Afroz, G. Acar, C. Diaz, and R. Greenstadt. A critical evaluation of website fingerprinting attacks. In *Proceedings of the 21st ACM Conference on Computer and Communications Security (CCS 2014)*, 2014.

79. C. V. Wright, S. E. Coull, and F. Monrose. Traffic morphing: an efficient defense against statistical traffic analysis. In *Network and Distributed System Security Symposium (NDSS)*, 2009.

80. A. Ibaida and I. Khalil. Wavelet-based ECG steganography for protecting patient confidential information in point-of-care systems. *IEEE Transactions on Biomedical Engineering*, 60(12):3322–3330, 2013.

81. M. A. Palacios, E. Benito-Pena, M. Manesse, A. D. Mazzeo, C. N. LaFratta, G. M. Whitesides, and D. R. Walt. Infobiology by printed arrays of microorganism colonies for timed and on-demand release of messages. *Proceedings of the National Academy of Sciences of the United States of America*, 108(40):16510–16514, 2011.

82. R. Lee, S. Schoen, P. Murphy, J. Alwen, and A. Huang. DocuColor tracking dot decoding guide. Technical Report, Xerox, 2005.

83. N. Aoki. A packet loss concealment technique for VoIP using steganography. *IEICE Transactions on Fundamentals of Electronics, Communications and Computer Sciences*, E86-A(8):2069–2072, 2003.

84. N. Aoki. VoIP packet loss concealment based on two-side pitch waveform replication technique using steganography. In *TENCON 2004, 2004 IEEE Region 10 Conference*, 2004.

85. N. Aoki. Potential of value-added speech communications by using steganography. In *Proceedings of the 3rd International Conference on International Information Hiding and Multimedia Signal Processing (IIH-MSP 2007), (IIH-MSP '07)*, Vol. 2, pp. 251–254. IEEE Computer Society, Washington, DC, 2007.

86. Y. Huang, J. Yuan, M. Chen, and B. Xiao. Key distribution over the covert communication based on VoIP. *Chinese Journal of Electronics*, 20(2):357–360, 2011.

87. I. Moskowitz, R. Newman, and P. Syverson. Quasi-anonymous channels. In *Proceedings of Communication, Network, and Information Security (CNIS)*, 2003.

88. J. Xu, J. Fan, M. Ammar, and S. Moon. Prefix-preserving IP address anonymization: measurement-based security evaluation and a new cryptography-based scheme. In *Proceedings of the 10th IEEE International Conference on Network Protocols*, pp. 280–289, November 2002.

89. S. J. Murdoch. Hot or not: revealing hidden services by their clock skew. In *Proceedings of the 13th ACM Conference on Computer and Communications Security (CCS '06)*, pp. 27–36. ACM, New York, USA, 2006.

90. R. deGraaf, J. Aycock, and M. J. Jacobson. Improved port knocking with strong authentication. In *Proceedings of the 21st Annual Computer Security Applications Conference (ACSAC '05)*, pp. 451–462. IEEE Computer Society, Washington, DC, 2005.

91. J. Kelley. Terror groups hide behind web encryption. *USA Today*, 2001.

92. D. Sieberg. Bin Laden exploits technology to suit his needs. *CNN*, 2001.

93. T. Kellen. Hiding in plain view: could steganography be a terrorist tool? *SANS Institute InfoSec Reading Room*, 2001.

94. R. Bryant. *Investigating Digital Crime*. Wiley, 2008.

95. H. Lau. The truth behind the Shady RAT. Technical Report, McAffe, 2011.

96. S. Adee. Spy vs. spy. *IEEE Spectrum*, August 2008.

97. S. Gallagher. How Al-Qaeda encrypted secret files within a porn video. *Wired*, 2011.

98. U.S. National Science and Technology Council, Interagency Working Group on Cyber Security and Information. Federal plan for cyber security and information assurance research and development. Technical Report, U.S. National Science and Technology Council, 2006.

99. W. Mazurczyk and L. Caviglione. Information hiding as a challenge for malware detection. *IEEE Security & Privacy*, 13(2):89–93, 2015.

100. R. Schlegel, K. Zhang, X. Zhou, M. Intwala, A. Kapadia, and X. Wang. Soundcomber: a stealthy and context-aware sound Trojan for smartphones. In *Proceedings of the 18th Annual Network and Distributed System Security Symposium (NDSS)*, February 2011.

101. L. Deshotels. Inaudible sound as a covert channel in mobile devices. In *Proceedings of the 8th USENIX Conference on Offensive Technologies (WOOT14)*, pp. 16–16. USENIX Association, Berkeley, CA, 2014.

102. M. Guri, G. Kedma, A. Kachlon, and Y. Elovici. AirHopper: bridging the air-gap between isolated networks and mobile phones using radio frequencies. *CoRR*, abs/1411.0237, 2014.

103. R. Hasan, N. Saxena, T. Haleviz, S. Zawoad, and D. Rinehart. Sensing-enabled channels for hard-to-detect command and control of mobile devices. In *Proceedings of the 8th ACM SIGSAC Symposium on Information, Computer and Communications Security (ASIA CCS '13)*, pp. 469–480. ACM, New York, 2013.

104. C. Mullaney. Morto worm sets a (DNS) record. Blog, August 2011.

105. B. Prince. Attackers hide communication within Linux backdoor. *Security Week*, 2013.

106. S. S. Response. Regin: top-tier espionage tool enables stealthy surveillance. Blog, November 2014.

107. D. Goodin. Duqu spawned by 'well-funded team of competent coders'—world's first known modular rootkit does steganography, too. *The Register*, 2011.

108. B. Bencsath, G. Pek, L. Buttyan, and M. Felegyhazi. Duqu: a Stuxnet-like malware found in the wild. Technical Report, Budapest University of Technology and Economics, 2011.

109. B. Stone-Gross. Malware analysis of the Lurk downloader. Blog, August 2014.

110. J. Gumban. Sunsets and cats can be hazardous to your online bank account. TrendLabs Blog, 2014.

111. S. Zander, G. Armitage, and P. Branch. A survey of covert channels and countermeasures in computer network protocols. *IEEE Communications Surveys & Tutorials*, 9(3):44–57, 2007.

112. S. Wendzel, S. Zander, B. Fechner, and C. Herdin. A pattern-based survey and categorization of network covert channel techniques. *ACM Computing Surveys (CSUR)*, 47(3):50:1-26, 2015.

113. S. Wendzel, V. Zwanger, S. Szlósarczyk, and M. Meier. Envisioning smart building botnets. I n *Proceedings of Sicherheit 2014*, pp. 319–329, 2014.

2

BACKGROUND CONCEPTS, DEFINITIONS, AND CLASSIFICATION

It is a riddle, wrapped in a mystery, inside an enigma; but perhaps there is a key...

—Winston Churchill

To understand the concept of *network steganography* let us consider the following real-life scenario: a crowded airport. Such an airport is like a router in a communication network and passengers are like packets—every day a lot of passengers are passing through an airport and they are traveling in many different directions. Also like packets in routers the passengers are "inspected" before entering a plane, for example, by using increasingly popular full-body scanners that are able to detect illegal objects on the passenger's body without making actual physical contact or without removing clothes.

However, even though this technology is continuously improved, still from time to time the security is breached and some illegal objects are smuggled into the plane. This is achieved by placing contraband in locations on the body where it is invisible on the body-scanner monitor.

To be able to deceive such a scanner the thorough knowledge of the device's inner functioning and limitations is required. This allows to identify potential vulnerabilities that can be exploited for illicit purposes.

Information Hiding in Communication Networks: Fundamentals, Mechanisms, Applications, and Countermeasures,
First Edition. Wojciech Mazurczyk, Steffen Wendzel, Sebastian Zander, Amir Houmansadr, and Krzysztof Szczypiorski.
© 2016 by The Institute of Electrical and Electronics Engineers, Inc. Published 2016 by John Wiley & Sons, Inc.

In general, the same principle is utilized in network steganography. To successfully transmit data in a covert manner, a solution to embed secret data into a network traffic flow must be found where the resulting modifications are not "visible" to network devices as well as end users.

2.1 CLASSIFICATION OF INFORMATION HIDING IN COMMUNICATION NETWORKS

In communication networks, there is a rich diversity of opportunities for information hiding that can be called "localizations" or hidden data carriers due to their increasing complexity and sophistication. Therefore, a great variety of information hiding techniques are potentially applicable. From the communications perspective, three main types of information can be subjected to hiding in networks (Figure 2.1):

- *Identities of communication parties*, hiding of sender and/or receiver identities to ensure their privacy, which can be achieved with a variety of anonymity techniques [1].
- *Communication process*, hiding the existence of the fact that data exchange is taking place can be accomplished using network steganography techniques, while traffic type obfuscation is used to conceal the characteristics of the transmission.

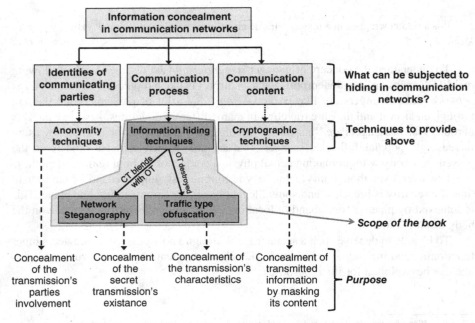

Figure 2.1. Classification of information concealment possibilities in communication networks.

- *Communication content*, cryptography-based techniques like encryption protect messages from disclosure to unauthorized parties thus concealing the content of exchanged messages. Other techniques in this group include, for example, scrambling of the user's content.

From the three groups of information concealment methods for communication networks provided above, only the techniques that hide the existence of the fact that data exchange is taking place are covered in this book (Figure 2.1). These methods can be further divided into *network steganography* and *traffic type obfuscation*, when we consider how they affect the overt transmission.

Network steganography hides data inside an overt communication in a way that minimizes the impact on the overt transmission(s) and thus it effectively conceals the existence of the covert transmission (CT).

On the other hand, traffic type obfuscation significantly changes or even completely alters an overt transmission (OT) to conceal the transmission's traffic characteristics by imitating another protocol's properties, for example by imitating the other protocol's packet frequency and/or size distributions.

Both groups of techniques will be described in detail in the next sections.

2.2 EVOLUTION OF INFORMATION HIDING TERMINOLOGY

As mentioned in Chapter 1, steganographic methods are the oldest information hiding techniques, dating back as far as ancient Greece [2]. Steganography, a word of Greek origin meaning "concealed writing," encompasses all concealing techniques that embed a secret message (steganogram) into a carrier in such a way that the carrier modification caused by the embedding of the steganogram must not be "noticeable" to anyone. The carrier is suitable for steganographic purposes if it is commonly used. The form of the carrier has evolved over time; historical carriers were wax tablets, human skin, or letters [2], but these days it is digital pictures, audio, text, or network protocols instead.

The last 15 years saw a very intensive research effort related to modern information hiding techniques. This was motivated by the industry business interested in DRM (Digital Rights Management) and the alleged utilization of steganographic methods by terrorists while planning the attacks carried out in the United States on the 11th of September, 2001 [3]. The latter was not a standalone case as in 2010, it was reported that a Russian spy ring had used digital picture steganography to leak classified information from the United States to Moscow [4].

In 1996 at the first Information Hiding (IH) Workshop held in Cambridge, United Kingdom, the following information hiding classification was agreed upon as presented in Figure 2.2 [2]. It is worth noting that before this event the term "information hiding" was formally used to describe a computer programming technique—encapsulation that was introduced by Parnas in 1972 [5].

In the classification presented by Petitcolas et. al. [2], the purpose of *steganography* was formulated as having a covert communication between two parties whose existence is unknown to a possible attacker; a successful attack consists of detecting the existence

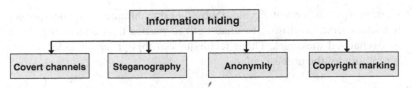

Figure 2.2. A historic classification of information hiding techniques. (Reproduced from [2] with permission of IEEE.)

of this communication (the definition is consistent with the definition from Simmons "prisoners' problem" [6]). *Copyright marking*, as opposed to steganography, has the additional requirement of robustness against possible removal or alteration attacks. In particular, copyright marks do not always need to be hidden. *Covert channels* were defined (supposedly after Lampson's definition [7]) as communication channels that were neither designed nor intended to transfer information at all. In [8] it is stated that covert channels are similar to steganography, but the cover-data is the whole behavior of a system. *Anonymity* encompasses techniques that hide the identity of the communicating parties thus ensuring their privacy.

The classification in [2] seems to be semantically inconsistent and misleading: (1) providing anonymity of communicating parties can be achieved by hiding information that identifies the communicating parties, but also, for example, by clever routing of data [8]; (2) for copyright marking, including watermarking, the main goal is to make the mark hard to remove, but whether it needs to be hidden or not depends on the application (for some applications it is important to have the mark clearly visible); (3) the distinction between steganography (and network protocol steganography in particular) and "covert channels" is not well grounded.

Obviously, the classification and definitions provided above were formulated considering the state-of-the-art at the time the IH workshop was held. However, in the last decade the information hiding field evolved very dynamically and therefore further clarifications and improvements are required. The developments in the scientific community over the last 19 years must be taken into account. Information hiding techniques have evolved, and this should be reflected by refining their classification and definitions. Providing coherent definitions and classifications that reflect the current state-of-the-art in information hiding is not an easy task, as for more than a decade there has been an ongoing debate between scientists in this research field. One of the important issues is the discussion of whether the relationship between steganography and covert channels exist, and whether both terms are synonyms or not.

In our opinion, the current distinction between steganography and covert channels, when used to describe information hiding in communication networks, is artificial and even misleading. Both terms do not describe separate hiding techniques, and their current distinction is simply due to the evolution of the hidden-data carrier. However, as we shall see both terms are related to each other.

The scientific community has used different terms, such as steganography, covert channels, or information hiding, to describe the process of concealing information in a digital environment. The variety of terms comes from the fact that the different terms had not been introduced at the same time and because their definitions evolved. Investigating the evolution of the definitions of "covert channel" and "steganography" will help to demonstrate that, currently, the distinction between both terms is artificial, especially in a communication networks environment. In communication networks, the hiding methods described by the two terms, *steganography* and *covert channels*, should be unified under a single term: *network steganography*. This does not mean that we want to discard the term covert channel. It is our view that network steganography techniques, as other steganography techniques, create covert (steganographic) channels for hidden communication, but such covert channels do not exist in communication networks without steganography (only the *possibility* for such channels exists *a priori*).

There are two widely known definitions of covert channels. Lampson [7], who coined the term in 1973, defined a covert channel as "channel, (...) not intended for information transfer at all." The U.S. Department of Defense (DoD) modified Lampson's definition to emphasize the malicious intentions with which a covert channel can be utilized in the so-called "Orange Book" in 1985 [9]. The DoD defined a covert channel as "any communication channel that can be exploited by a process to transfer information in a manner that violates the system's security policy", which implied certain applications of the covert channel. It is worth noting that both of these definitions do not imply what techniques are utilized to create the hidden communication channel.

During the last 19 years, the covert channel's definition blurred and currently it is often used to describe particular information hiding methods, for example, the hiding of information in unused/optional fields, rather than the associated communication channels. Moreover, in many papers (e.g., in [2] and [10]) steganography is intentionally distinguished from covert channels. Covert channels, as their name implies, are in general used for communication, but in theory one could also use them to store information, for example, in network traffic traces or other network-related log files stored on hard disk. Steganography may be used to store information, for example, one can embed secret data into a digital image and store it on a hard disk to hide the secret information from everyone, but of course it is also often used for communication. In fact ancient steganographic techniques were often used for communication, that is, hidden messages were concealed in innocent looking carriers (see Chapter 1 for examples) to convey information to a receiving party aware of the steganographic procedure.

For example, let us consider again a famous steganographic method described by Herodotus [2] where a trusted slave's head was shaved and then tattooed with a secret message about the planned revolt against the Persians. After his hair had regrown the slave was sent with this message to the Ionian city of Miletus. The goal of this steganographic method as well as many others that followed was to conceal the hidden communication by hiding secret information (in form of a tattoo) in a cover (carrier) of this data (tattooed human skin). Thus, the main aim of the historical steganographic techniques was to create covert channels (in uni- or bidirectional manner). In general, they were used not to store information concealed from everyone but to hide the information transfer.

In our opinion, the main difference between the terms steganography and covert channels is merely the type of the carrier (cover) used; for steganography it is digital media like images, audio, and video files whereas for covert channels these are network protocols. Both terms exist because the communication methods used by people evolved from messengers, letters, and telephones to communication networks and accordingly the steganographic techniques evolved with the available communication methods. However, information hiding techniques that utilize network protocols should be treated as an evolutionary step of the carrier rather than some new phenomenon.

We propose to broaden the definition of the term covert channel so it describes any covert channel created by steganographic techniques, and we propose to use the term network steganography to describe the *methods* used for creating covert channels in communication networks. Our approach has the advantage that it provides the long-needed clarification, but at the same time it is backwards-compatible with the existing literature. While the term network steganography should be used to describe a hiding technique in communication networks, it is not wrong to refer to the resulting covert channel.

2.3 NETWORK STEGANOGRAPHY: DEFINITIONS, CLASSIFICATION AND CHARACTERISTIC FEATURES

We propose the following classification of modern steganography techniques (not limited to communication networks) as presented in Figure 2.3.

The naming convention is *carrier-based*. This means that if network traffic is used as a carrier (the majority of so called covert channels [10]) then this is network steganography. When digital content, like images, audio, or video files, is used as a carrier then this is image, audio, or video steganography [11]. In Figure 2.3, these techniques are aggregated into a single digital media steganography category. The previously named subliminal channels are renamed to cryptosystem steganography to ensure consistency. Other types of steganography mentioned in Figure 2.3 include, but are not limited to, physical steganography, where the most notable example is the previously mentioned information hiding by tattooing human skin, and filesystem steganography, where a filesystem's features are exploited for secret data concealment. Other types of steganography mentioned in Chapter 1 can be easily fitted into the proposed classification.

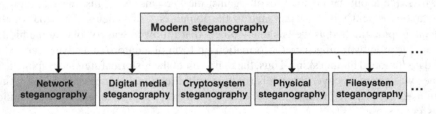

Figure 2.3. Classification of modern steganography techniques and scope of network steganography.

The main aim of network steganography is to hide secret data in the normal transmissions of users without significantly altering the carrier used. The scope of network steganography is limited to all information hiding techniques that:

- can be applied to the network traffic to hide the exchange of data by creating covert communication channel(s).
- are inseparably bound to the transmission process (carrier).
- do not significantly alter the carrier.

It must be emphasized that the main difference between "classic" steganography and network steganography is that the first relied on fooling human senses and the latter mainly tries to deceive other network devices (intermediate network nodes or end systems). For a third-party observer who is not aware of the steganographic procedure, the exchange of steganograms remains hidden. This is possible because the embedding of hidden data into a chosen carrier is "invisible" for parties not involved in the steganographic communication. Thus, the secret data is hidden inside the carrier, and the fact that the secret data is exchanged is also concealed.

In network steganography, a *carrier* is one or more overt traffic flows that pass between a covert/steganogram sender and a covert/steganogram receiver(s). A carrier can be multidimensional, that is, it offers many opportunities (places) for information hiding (called *subcarriers*). A *subcarrier* is defined as a "place" or a timing of "events" in a carrier (e.g., a header field, padding, or an intended sequence of packets) where secret information can be hidden using a single steganographic technique (Figure 2.4). Typically a subcarrier takes the form of a storage or a timing covert channel.

Subcarriers are also referred to as "cover protocols" [12] but the definition of cover protocols is limited to storage areas in network protocol headers and also limited to the context of transferring control information in network covert channels whereas the term "subcarrier" can be applied in a broader meaning (see the book's Glossary).

As visualized in Figure 2.4 for an example VoIP connection, multiple subcarriers can be used simultaneously within a single overt network flow. Another example that illustrates the same concept is an HTTP flow used as steganographic carrier with two subcarriers, one which embeds secret data into the HTTP User Agent field and another that embeds secret data into the IP ToS (Type of Service) field. Both subcarriers could also be located within the same layer of the TCP/IP stack.

To hide secret data in a subcarrier, it can be necessary to create space for its placement. For example, a covert sender can create a new IPv4 option or a new IPv6 destination option and then embed the secret data into this allocated space.

The most favorable carriers for secret messages in communication networks must have two features:

- they should be popular, that is, the use of such carriers should not be considered as an anomaly. The more such carriers are present in a network, the easier it is to mask the existence of hidden communication.

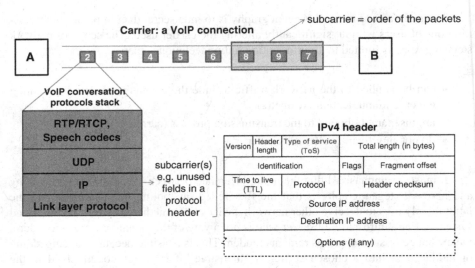

Figure 2.4. An example of carrier and subcarriers based on VoIP connection example. (Reproduced from [13] with permission of Wiley.)

- modification of the carrier related to embedding of the steganogram should not be "visible" to parties unaware of the steganographic procedure.

Steganography relies on three characteristics of communications in current communication networks. First, a communication channel is not perfect. Errors and network anomalies, such as corrupted, lost, or reordered packets, are a natural phenomenon and, thus, it is possible to embed information by mimicking them. Second, most network protocols specify fields or messages that are not used in all situations. This "surplus" can be used for embedding of secret data, if this does not degrade the carrier. Third, not every protocol is completely defined and "semantic overloading" is possible. Most of the specifications permit some amount of freedom in the implementation, and this can be utilized for steganographic purposes, for example, HTTP header fields can be lower or upper case and secret data can be encoded via manipulating the case. Designing steganography-free protocols is hard and often practically impossible without unreasonably limiting the functionality or extensibility of protocols. Hence, network steganography can be achieved even with the simplest of communication protocols. However, more complex protocols usually offer more opportunities for sophisticated information hiding methods.

Considering the above definitions of carrier and subcarrier it is also important to emphasize that network steganography can be applied to *single* or *multiple* traffic flows. The example in Figure 2.4 presents the concept of single-flow network steganography, while Figure 2.5 visualizes the concept of multiple-flows network steganography. In the latter case, the hidden data bits are distributed among many network flows. For instance, Figure 2.5 shows how binary "1" and "0" can be encoded into the sequence of packets from three different TCP connections.

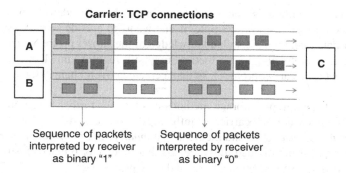

Figure 2.5. Multiple-flows steganography example—sending secret data that is distributed over a number of traffic flows. (Reproduced from [13] with permission of Wiley.)

We propose the following network steganography classification illustrated in Figure 2.6. It consists of three levels. Firstly, network steganography can be divided into *timing* and *storage* methods based on how the secret data is encoded into the carrier (which is consistent with previously introduced covert channels classifications). Of course by combining the features of both methods hybrid solutions are also possible.

Then each of the two groups can be divided further. If we take into account whether timing methods depend on the underlying protocol we can distinguish between *protocol aware* and *protocol agnostic* methods. Protocol-aware timing methods require understanding of the carrier protocol—they utilize its characteristic features for hidden data exchange. For example, if a method exploits the fragmentation mechanism of a certain protocol, it needs to take into account how exactly this protocol's fragmentation mechanism works. On the other hand protocol-agnostic techniques can be applied blindly to

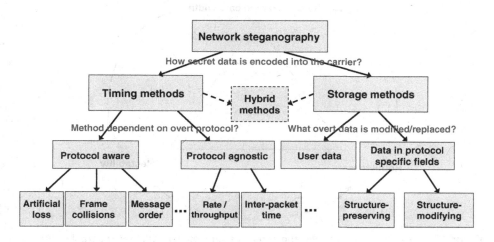

Figure 2.6. Network steganography methods classification.

the selected carrier without in-depth understanding of the utilized protocol mechanisms and their specific features. For example, if one wants to encode secret data in interarrival time or data rate, then a detailed knowledge of the carrier protocol is not required. However, even a protocol-agnostic method may still require knowledge of the traffic characteristics of the carrier protocol, so that a created covert channel looks like normal traffic.

Network steganography storage methods can be split further based on what overt data is modified (which subcarrier is influenced). We can distinguish two groups: methods that *modify user data* and methods that *modify data in protocol specific fields*. The latter can be further divided into methods that modify or replace existing protocol data and those that insert additional nonmandatory protocol data.

Each network steganography method can be characterized by three features. First, *steganographic bandwidth* describes how much secret data one is able to send per time unit. Second, *undetectability* is defined as an inability of an adversary to detect a steganogram inside a carrier. The most popular way to detect a steganogram is to analyze statistical properties of the captured data and compare them to the typical properties of that carrier. Third, *robustness* is defined as the amount of alteration a steganogram can withstand without destroying the secret data. For each network steganography method, there is always a trade-off between maximizing steganographic bandwidth and still remaining undetected (and retaining an acceptable level of robustness). A user can utilize a method naively and send as much secret data as possible, but it simultaneously raises the risk of disclosure. Therefore, he or she must purposely limit the steganographic bandwidth in order to avoid detection. The relationships between these three features are typically described by a magic triangle as proposed by Fridrich [14] and illustrated in Figure 2.7.

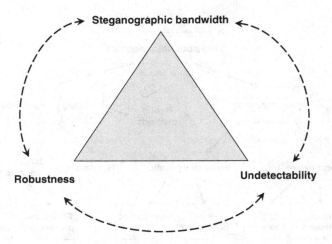

Figure 2.7. Relationship between the three features of network steganography. (Reproduced from [15] with permission of ACM.)

Often it is also useful to calculate a *steganographic cost* [13]. It describes the degree of degradation of the carrier caused by the steganogram insertion. The steganographic cost depends on the type of the carrier utilized, and if it becomes excessive, it leads to easier detection of the steganographic method. The cost indicates the degradation or distortion of the carrier caused by the application of a steganographic method. In digital media steganography, that is, for hiding secret data in digital image, audio, or video, the parameters MSE (Mean-Square Error) or PSNR (Peak Signal-to-Noise Ratio) are typically utilized for this purpose. However, these parameters cannot be applied to dynamic, diverse carriers like network connections. Instead, other parameters must be used. For example, if the steganographic method uses voice packets from IP telephony as a carrier, then the steganographic cost can be expressed in conversation degradation. If the carrier is a certain protocol field, then the cost can be expressed as a potential loss in protocol functionality, etc. It is also possible that an information hiding method introduces steganographic cost that can be experienced in two different dimensions, for example, it introduces voice quality degradation as well as it adds additional delays to the overt traffic. If we include steganographic cost as a feature of a network steganography method, then the relationships with the remaining features is as illustrated in Figure 2.8. When a steganographer wants to maximize steganographic bandwidth it typically results in increased steganographic cost, higher detetctability, and lower robustness.

Therefore, in general, it can be concluded that the steganographic cost affects detectability and may be responsible for the reduction of the carrier's functionality or the reduction of the carrier's performance (e.g., it results in increased transmission times or increased usage of resources). The relationship between steganographic cost and detectability is explained in Figure 2.9. One can imagine a steganographic cost as a "zip" as it provides a view on how exactly the carrier is affected by applying a steganographic method. On the other hand, undetectability can be imagined as an "on/off switch". When a given level of steganographic cost (SC_T) is exceeded, the steganographic method becomes detectable with a probability greater than 50% ("flip a coin" chance of detection) up to the point where the detection is trivial ($SC_{D=100\%}$).

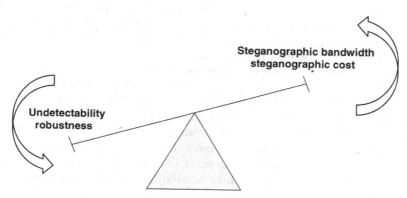

Figure 2.8. Relationship between the features of network steganography with steganographic cost included. (Reproduced from [13] with permission of Wiley.)

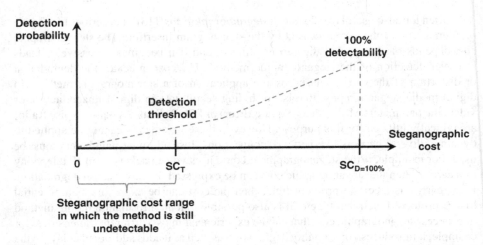

Figure 2.9. Relationship between steganographic cost and undetectability. (Reproduced from [13] with permission of Wiley.)

2.4 TRAFFIC TYPE OBFUSCATION: DEFINITIONS, CLASSIFICATION AND CHARACTERISTIC FEATURES

Traffic type obfuscation hides the nature of network traffic flows by obfuscating the underlying network protocol. That is, traffic type obfuscation modifies the patterns and contents of network traffic between two network entities so that a third entity is not able to reliably identify the type of their communication, that is, the network protocol. We classify the existing work on traffic type obfuscation into two broad categories: *traffic de-identification* and *traffic impersonation*. Traffic de-identification manipulates network traffic so that the underlying network protocol is hidden. For instance, network packet headers are "encrypted" to conceal protocol identifier contents, thus obfuscating the underlying network protocol. On the other hand, traffic impersonation manipulates traffic so that it not only hides the underlying network protocol, but also pretends to be using another protocol, for example, a target protocol. For instance, a BitTorrent client may alter its traffic to look like HTTP traffic.

There are different reasons why someone would want to conceal the network protocol through traffic obfuscation. First, it can be used to bypass network firewalls that disallow certain network protocols. For instance, many ISPs filter out peer-to-peer file sharing protocols like BitTorrent, where an efficient evasion is to obfuscate the protocol through traffic type obfuscation. Second, it can be used to conceal the existence of a type of communication. For example, the users of censorship circumvention tools may need to hide their use of circumvention systems, like Tor [8]; this requires them to hide the Tor protocol, for example, by morphing their traffic to look like a legitimate network protocol [16].

We classify all techniques as traffic type obfuscation that make the network protocol unidentifiable with respect to

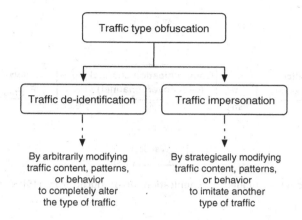

Figure 2.10. Traffic type obfuscation techniques classification.

- contents of network packets,
- statistical patterns of network packets, such as packet timing and sizes, and
- protocol behavior, such as dynamic reaction to certain events.

As noted above, traffic type obfuscation shares intrinsic similarities with network steganography. Consequently, most of the terminologies and models used for traffic obfuscation are borrowed from the older, more established area of network steganography. More specifically, traffic type obfuscation is similar to network steganography, described in previous section, in that both aim to hide the existence of a covert communication. However, they differ in two main ways. First, in network steganography the overt channel is not significantly altered. By contrast, in traffic type obfuscation the contents of the carrier are "destroyed" by replacing it with the covert content. As a result, traffic type obfuscation techniques are able to provide higher capacities for covert communications. Second, traffic type obfuscation and network steganography differ in the application scenarios that they can be used for. Traffic type obfuscation can only be used when the carrier flow is generated by covert parties (covert senders and receivers). In contrast, network steganography can be applied on traffic flows regardless of who has generated them.

As described before, the main objective of network steganography is to communicate the highest amount of secret information without being detected. In traffic type obfuscation, however, the objective is to conceal the type of the network traffic. As a result, the features such as "bandwidth", "robustness", and "cost" are not applicable to traffic type obfuscation. Instead, a type obfuscation scheme should be "unobservable": third parties should not be able to (1) detect that the traffic has been obfuscated, and (2) determine the original type of network traffic. The feature "unobservability" is equivalent to the feature "undetectability" used in the network steganography context.

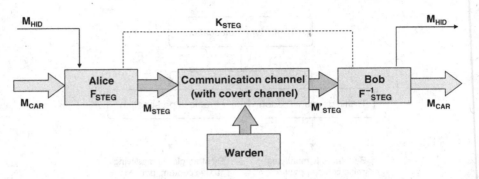

Figure 2.11. Model for hidden communication. (Reproduced from [15] with permission of ACM.)

2.5 HIDDEN COMMUNICATION MODEL AND COMMUNICATION SCENARIOS

The state-of-the-art communication model for steganography [2] and covert channels [10] is the same model. It is the famous "prisoners' problem", which was first formulated by Simmons [6] in 1983 (Figure 2.11). In this model, Alice and Bob are two prisoners that are trying to prepare an escape plan. The problem is that their communication is monitored by a Warden. If the Warden identifies any conspiracy, he will put Alice and Bob into solitary confinement making an escape impossible. So, Alice and Bob must find a way to exchange hidden messages for their escape plan to succeed. The solution is to use steganography. By concealing the hidden message (M_{HID}) in an innocent looking carrier (M_{CAR}), it is possible to generate a modified carrier (M_{STEG}) that will raise no suspicion while traversing through the communication channel. For Alice and Bob, the communication channel is also a covert channel that was created using a steganographic method.

Therefore, if there is no difference in the communication model it could be concluded that the term covert channel was used as another term for steganography (this fact was also observed by other researchers in the field, for example, by Fridrich in [17]). However, we are convinced that the term covert channel cannot be used as a synonym for the term steganography. We believe that the term covert channel should be used to describe the communication channel that is created with the use of a steganographic method and this is consistent with Lampson's original definition of covert channels (see above). Thus, to be able to create a covert channel through which hidden data is exchanged, the sender (F_{STEG} in Figure 2.11) and the receiver (F_{STEG}^{-1} in Figure 2.11) must always utilize a steganographic method.

For steganographic methods, it is usually assumed that there exists a secret stego-key (K_{STEG} in Figure 2.11) that is a shared secret between Alice and Bob. In network steganography, a knowledge of *how the information is hidden* is the stego-key. In many cases, network steganography achieves security through obscurity; only if a steganographic technique is unknown to the Warden, it can be used to exchange hidden data securely. However, some techniques are secure even if their existence is known, as long as the steganographic parameters are unknown to the Warden (e.g., properly encoded TCP ISN steganography [18]). Hence, the stego-key is a combination of the steganography technique and its parameters used by Alice and Bob.

Apart from the stego-key everything else can be known to the Warden. In particular the Warden is aware that Alice and Bob can utilize hidden communication, *can* know all other existing steganographic methods (unrelated to the stego-key), and is able to try to detect and/or interrupt the hidden communication. We refer to a *passive* Warden if the Warden tries to detect a steganographic communication or if the Warden tries to determine the involvement of a party into the steganographic communication. An *active* Warden, on the other hand, tries to eliminate or manipulate the steganographic communication. A *malicious* Warden can additionally alter Alice's or Bob's messages or even introduce fake messages into the communication [19].

Let us consider the possible hidden communication scenarios (S1-S4 in Figure 2.12), as they greatly influence the detection possibilities for the warden. In general, there are three possible localizations for a warden (denoted in Figure 2.12 as W1-W3). A node that performs steganalysis can be placed near the sender, or receiver of the overt

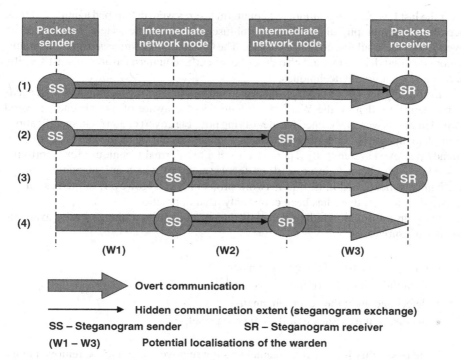

Figure 2.12. Hidden communication scenarios and potential localizations of the warden. (Reproduced from [15] with permission of ACM.)

communication or at some intermediate node. Moreover, the warden can monitor network traffic in single (*centralized* warden) or multiple locations (*distributed* warden). In general, the localization and number of locations in which the warden is able to inspect traffic influences the warden's effectiveness.

The communication model is slightly different in the case of traffic type obfuscation. Unlike network steganography, which aims to send covert data, traffic type obfuscation tries to conceal the type of network traffic. Consider two entities Alice and Bob who aim to communicate in the presences of a warden. Suppose that the warden forbids any communication of type T1. Suppose that Alice and Bob want to communicate using traffic of type T1. To be able to do so, they need to obfuscate the type of traffic, for example, by mimicking an allowed type of traffic, T2.

Moreover, unlike network steganography, traffic type obfuscation can only be applied when the traffic flows are generated by the covert entities. That is, traffic type obfuscation entities only obfuscate the network flows generated by themselves. As a result, the only communication scenario used in type obfuscation is the model (1) from Figure 2.12.

2.6   INFORMATION HIDING COUNTERMEASURES MODELS

Over the last few decades a number of countermeasures were developed against different network steganography and traffic type obfuscation techniques. Here, we provide an overview of the available countermeasures. The Chapter 8 *"Countermeasures"* contains a more detailed discussion and also describes specific countermeasures against specific network steganography techniques.

Before any action can be taken against an information hiding technique, it first needs to be identified, that is, the Warden needs to become aware of its existence. Several formal methods were developed for identifying possible covert channels in specifications or implementations of operating systems or applications during the design phase or in an already deployed system [20]. There also exist a few formal techniques for identifying possible covert channels in network protocols, many of which are adaptations of the mechanisms mentioned in [20] to network protocols [21]. However, to the best of our knowledge none of these has been commonly used in practice.

Once an information hiding technique has been identified, the generally available countermeasures are

- eliminate the use of the covert channel,
- limit the bandwidth of the covert channel,
- detect and audit the covert channel,
- document the existence of the covert channel.

If the possibility for a covert channel was not removed or cannot be removed in the protocol design or implementation, the next best option is to *eliminate the use* of the channel, because even low-capacity channels could be successfully exploited. The direct approach is to block or eliminate network protocols that enable network steganography. However, this approach is often impractical, since there are too many possibilities for hiding information in network protocols. The most common approach to prevent the use of network steganographic techniques and eliminate the resulting covert channels is *traffic normalization*.

Traffic normalizers (e.g., Snort [22]) remove semantic ambiguities by normalizing protocol header fields or timing behaviors of network traffic passing through the normalizer. For example, a normalizer can always set unused, reserved, and padding bits to zero, thus removing the possibility to hide information in these bits. Traffic normalization can only be applied to all network traffic ("blind" normalization) if the normalization is transparent (it does not affect the traffic significantly). In practice, the use of normalizers may result in unwanted side-effects, since the normalization of header fields often means setting header fields to default values, that is, these fields are not usable anymore. If accurate detection methods exist, detected covert channels can be eliminated using targeted normalization or even disruptive measures, for example, the carrier traffic could simply be blocked.

The removal of all covert channels leads to very inefficient systems, since it typically means replacing automated procedures with manual procedures [23]. Furthermore, covert channels based on the modulation of message characteristics are inherent in distributed systems, such as computer networks. Therefore, we and many other experts in the field believe that covert channels cannot all be completely eliminated [24]. This is also acknowledged by the security standards. For example, the Orange Book treats covert channels with capacities of less than one bit per second as acceptable in many scenarios [9].

If a channel cannot be eliminated its *capacity should be reduced* below an acceptable limit. What is an acceptable capacity depends on the amount of information leakage that is critical. For example, if the covert channel capacity is so small that classified information cannot be leaked before it is outdated, then the channel is tolerable. Limiting the channel capacity is often problematic in reality, because it means slowing down protocol mechanisms or introducing noise, which both limit the performance of the protocol. However, some limitation techniques showed good results in the literature as well as in practice, especially in the case of network steganographic timing methods that are hard to prevent. For example, the PUMP method [25] is very effective in limiting (practically eliminating) timing channels in the flow of acknowledgment messages from a receiver to a sender needed for a reliable data transfer.

Covert channels that *cannot be eliminated or limited should be audited*, which requires reliable detection methods. Auditing acts as deterrence to possible users and also allows taking actions against actual users. (If the traffic data used for auditing can be obtained before traffic normalization takes place, one may want to audit the use of all covert channels.) The detection of covert channels is commonly based on statistical approaches or machine learning (ML) techniques [10]. In both cases the behavior of actual observed network traffic is compared against the known assumed behavior of "normal" traffic and (if known) the assumed behavior of carrier traffic with covert channels. The behavior is measured in the form of characteristics (also called *features* in ML terminology) computed for the actual observed traffic and traffic previously used to determine a *decision threshold* (or *decision boundary*) between normal traffic and traffic with steganography (training of the detection system).

With statistical approaches the decision boundary usually must be determined manually (by a human) during the training, and for some approaches proposed in the literature is not very clear how to choose an effective boundary. In contrast, ML methods determine the decision boundary automatically during the training phase. A disadvantage with some ML techniques is that the decision boundary can be very hard to interpret—effectively making the detection system a black box.

Supervised ML techniques require training with examples for both *classes* (normal traffic and traffic with covert channels). *Unsupervised ML* techniques or *anomaly detection* methods are usually trained with examples of normal traffic only. Supervised ML techniques are often more accurate, but also are less robust. Without proper retraining they can easily fail if the characteristics of one of the classes change, for example, due to a change of the covert channel encoding or a change of the normal protocol operations. Unsupervised techniques are often less accurate, but are more robust against changes in traffic characteristics. With all detection methods it is important to avoid creating a detector that performs very well for the training data, but does not perform well for the actual observed traffic (*overfitting* on the training data).

All known covert channels, except channels with capacities that are too low to be significant, should at least be documented (for example, in the specification of network protocols). This makes everybody aware of their existence and potential threat, and it also deters potential users, since many steganographic techniques only provide security by obscurity.

As described before, the main objective of traffic type obfuscation is to conceal the type of network traffic, that is, the network protocol being used. Consequently, any countermeasures against traffic type obfuscation will have one or both of the following two objectives: First, to detect network flows that try to hide their actual network protocols, and, second, to identify the actual network protocols of those network flows. If the actual protocol can be identified further actions can be taken in accordance with existing security policies, for example, the obfuscated traffic can be blocked.

Countermeasures against traffic type obfuscation can be classified into three categories: content-based countermeasures [26], which look for content discrepancies in the obfuscated traffic, pattern-based countermeasures [27,28], which use statistical tools to identify obfuscated traffic based on their communication patterns (for example, packet sizes and timings), and protocol-based countermeasures [26,29] that analyze the protocol behavior of traffic, for example, by investigating their reaction to certain network conditions.

2.7 SUMMARY

This chapter discussed information concealment possibilities in communication networks, dividing potential methods into those that hide the communicating parties' identities, those that hide the communication process and those that hide the data exchanged. However, the main focus was on techniques that enable covert communication, that is, network steganography and traffic type obfuscation. The main difference between these two is how the overt communication is treated. In network steganography, the overt traffic is modified very slightly but in traffic type obfuscation the overt traffic is completely altered. For both of these subdisciplines of information hiding, we characterized their main effectiveness features, communication models, and typical usage scenarios. Finally, the countermeasures were briefly discussed.

REFERENCES

1. G. Danezis and C. Diaz. A survey of anonymous communication channels. Technical Report MSR-TR-2008-35, Microsoft Research, 2008.

2. F. A. Petitcolas, R. Anderson, and M. Kuhn. Information hiding-a survey. *Proceedings of the IEEE*, 87(7):1062–1078, 1999.

3. D. Sieberg. Bin laden exploits technology to suit his needs. *CNN*, 2001.

4. N. Shachtman. Fbi: Spies hid secret messages on public websites. *Wired*, 2010.

5. D. L. Parnas. On the criteria to be used in decomposing systems into modules. *Communications of the ACM*, 15(12):1053–1058, 1972.

6. G. J. Simmons. The prisoners' problem and the subliminal channel. In D. Chaum, editor, *Advances in Cryptology*, pp. 51–67. Springer, New York, 1984.

7. B. W. Lampson. A note on the confinement problem. *Communications of the ACM*, 16(10):613–615, 1973.

8. R. Dingledine, N. Mathewson, and P. Syverson.Tor: The Second-generation Onion Router. In *USENIX Security*, 2004.

9. DoD. Orange book, trusted computer system evaluation criteria, tech. rep. dod 5200.28-std. Technical report, DoD, 1985.

10. S. Zander, G. Armitage, and P. Branch. A survey of covert channels and countermeasures in computer network protocols. *Communications Surveys Tutorials, IEEE*, 9(3):44–57, 2007.

11. W. Bender, D. Gruhl, N. Morimoto, and A. Lu. Techniques for data hiding. *IBM Systems Journal*, 35(3.4):313–336, 1996.

12. S. Wendzel and J. Keller. Systematic engineering of control protocols for covert channels. In B. De Decker and D. Chadwick, editors, *Communications and Multimedia Security*, volume 7394 of *Lecture Notes in Computer Science*, pp. 131–144. Springer, Berlin, 2012.

13. W. Mazurczyk, S. Wendzel, I. A. Villares, and K. Szczypiorski. On importance of steganographic cost for network steganography. *International Journal of Security and Communication Networks*, 2014.

14. J. Fridrich. Applications of data hiding in digital images. In *Proceedings of the Fifth International Symposium on Signal Processing and Its Applications, 1999 (ISSPA '99)*. Vol. 1, pp. 1–9, 1999.

15. W. Mazurczyk. VoIP steganography and its detection: a survey. *ACM Computing Surveys*, 46(2):20:1–20:21, 2013.

16. H. Moghaddam, B. Li, M. Derakhshani, and I. Goldberg.SkypeMorph: protocol Obfuscation for Tor Bridges. In *CCS*, 2012.

17. J. Fridrich. *Steganography in Digital Media: Principles, Algorithms, and Applications*. Cambridge University Press, New York, NY, 1st edition, 2009.

18. S. J. Murdoch and S. Lewis. Embedding covert channels into tcp/ip. In *Proceedings of the 7th International Conference on Information Hiding* (IH'05), pp. 247–261. Springer-Verlag, Berlin, 2005.

19. S. Craver. On public-key steganography in the presence of an active warden. In *Information Hiding*, Vol. 1525 of *Lecture Notes in Computer Science*, pp. 355–368. Springer, Berlin, 1998.

20. V. Gligor. A guide to understanding covert channel analysis of trusted systems. Technical report, National Computer Security Center, Technical Report NCSC-TG-030, 1993.

21. A. L. Donaldson, J. McHugh, and K. A. Nyberg. Covert channels in trusted lans. In *Proceeding of 11th NBS/NCSC National Computer Security Conference*, pp. 226–232, 1988.

22. Snort Project. Snort user's manual 2.9.3, 2012.

23. N. E. Proctor and P. G. Neumann. Architectural implications of covert channels. In *Proceeding of 15th National Computer Security Conference*, pp. 28–43, 1992.

24. I. Moskowitz and M. Kang. Covert channels-here to stay? In *Safety, Reliability, Fault Tolerance, Concurrency and Real Time, Security: Proceedings of the 9th Annual Conference on Computer Assurance (COMPASS '94)*, pp. 235–243, June 1994.

25. M. H. Kang and I. S. Moskowitz. A pump for rapid, reliable, secure communication. In *Proceedings of the 1st ACM Conference on Computer and Communications Security* (CCS '93), pp. 119–129. ACM, New York, NY, 1993.

26. A. Houmansadr, C. Brubaker, and V. Shmatikov. The Parrot is Dead: observing unobservable network communications. In *34th IEEE Symposium on Security and Privacy,* Oakland, CA, 2013.

27. X. Cai, X. C. Zhang, B. Joshi, and R. Johnson. Touching from a distance: website fingerprinting attacks and defenses. In *Proceedings of the 2012 ACM Conference on Computer and Communications Security*, pp. 605–616. ACM, 2012.

28. K. P. Dyer, S. E. Coull, T. Ristenpart, and T. Shrimpton. Peek-a-Boo, I still see you: why efficient traffic analysis countermeasures fail. In *Proceeding of the 2012 IEEE Symposium on Security and Privacy (SP)*, pp. 332–346. IEEE, 2012.

29. J. Geddes, M. Schuchard, and N. Hopper. Cover your ACKs: pitfalls of covert channel censorship circumvention. In *Proceedings of the 2013 ACM SIGSAC conference on Computer & Communications Security (CCS)*, pp. 361–372. ACM, 2013.

3

NETWORK STEGANOGRAPHY

The best way of keeping a secret is to pretend there isn't one.

—Margaret Atwood, The Blind Assassin: A Novel

The aim of network steganography is to hide the communication process, or in other words, to hide the existence of a data exchange (see Chapter 2). Network steganography techniques use overt network traffic as a carrier for the secret data. They hide the secret data inside the overt communication such that the impact on the overt transmission(s) is minimal, and thus the covert transmission is concealed effectively. The covert transmission is commonly referred to as a covert channel.

Network steganography techniques can be classified into *storage* and *timing* methods based on how the secret data are encoded into the carrier (see Chapter 2). Storage methods hide information by modifying protocol fields, such as unused bits of a header. Timing methods hide information in the timing of protocol messages or packets. *Hybrid* solutions are methods that combine the features of timing *and* storage methods.

A large number of hiding methods were introduced within the last few decades, which is linked to the following drawbacks: *First*, a coverage of all known hiding methods would exceed the page limit of this book. *Second*, various hiding methods are

Information Hiding in Communication Networks: Fundamentals, Mechanisms, Applications, and Countermeasures,
First Edition. Wojciech Mazurczyk, Steffen Wendzel, Sebastian Zander, Amir Houmansadr, and Krzysztof Szczypiorski.
© 2016 by The Institute of Electrical and Electronics Engineers, Inc. Published 2016 by John Wiley & Sons, Inc.

based on the same or similar concepts and a coverage of all methods would thus lead to redundancy. *Third*, a simple coverage of all hiding methods would not necessarily comprise an optimal structure or taxonomy of hiding methods.

Wendzel et al. provide an organized view on hiding methods [1]. They propose a taxonomy of so-called *hiding patterns*, where each hiding pattern is a unified and generic description of a particular hiding method.

A hiding pattern's description is written in an abstract manner so that one pattern can be used to describe multiple hiding techniques at the same time. For instance, "modulate the least significant bits of a protocol field" is a very brief description for many published hiding methods that utilize the least significant bits of fields in arbitrary network protocols.

An advantage of the pattern-based approach in [1] is that researchers who want to publish a new hiding method now do not need to analyze all the known hiding methods anymore to verify whether their own method is entirely novel. Instead, they only need to compare their method with the existing hiding patterns that are similar. Each pattern points to its related literature. With the related literature, the researchers can motivate their new method by highlighting the exact differences to previous work (e.g., higher level of stealthiness or higher channel capacity). This reduces the required time for research and may result in fewer reinventions.

First, we describe hiding patterns based on protocol modifications. Next, we introduce hiding patterns based on the timing of protocol messages. Finally, we discuss hybrid methods, which can be seen as parallel applications of at least two hiding patterns.

3.1 HIDING INFORMATION IN PROTOCOL MODIFICATIONS

In Chapter 2, we divided storage methods into those using user data and those hiding data in specific fields of protocols. For the latter case, data can either be hidden in a structure-preserving or a structure-modifying manner (cf. Figure 2.6). The extended storage method taxonomy of [1] is shown in Figure 3.1.

Like in the classification of Chapter 2, patterns do either hide in user data (*payload*) or in protocol fields (*non-payload*). There are two types of non-payload patterns:

- *Structure-modifying* patterns alter the structure of a *protocol data unit* (PDU), for example, by changing its size.
- *Structure preserving* patterns keep the given PDU structure and modify the different types of PDU header fields (*attributes*), for example, a reserved bit.

We will now discuss all patterns individually in the seen order of Figure 3.1. While for each pattern selected hiding methods are explained, it is important to know that many more hiding methods can be found for each of the patterns in the literature [1].

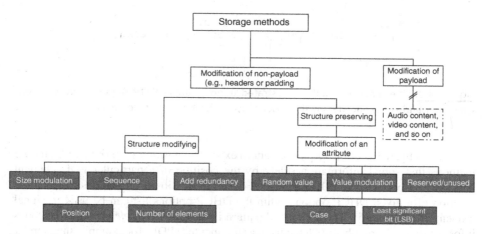

Figure 3.1. Taxonomy for storage methods as patterns shaded. (Reproduced from [1] with permission of ACM.)

3.1.1 Size Modulation of Protocol Data Units

Hiding techniques can signal information by altering the size of a *protocol data unit* and through the size of a particular header element within a PDU. For instance, Murdoch and Lewis [2] published work in which the size of IP fragments is modulated and Girling modulates the data block length in LAN frames [3]. The pattern's functioning is visualized in Figure 3.2.

3.1.2 Sequence Modulation in PDUs

The sequence pattern comprises all hiding methods that alter the order of a PDU's elements. Such approaches can easily be designed for protocols with highly dynamic header structure such as HTTP. Dyatlov and Castro show that the order of different header lines in the HTTP header can encode hidden information [4]. The method is visualized in Figure 3.3. Rios et al. show that it is also feasible to manipulate the order of the DHCP options [5], and Zou et al. demonstrate the same method by changing the sequence of FTP commands [6].

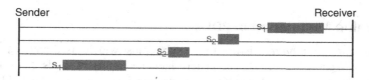

Figure 3.2. Illustration of the size modulation pattern: PDUs of different size are transmitted between sender and receiver to encode symbols s_1 and s_2.

```
GET HTTP/1.1                        GET HTTP/1.1
Host: mywebsite.xyz                 Host: mywebsite.xyz
User-Agent: MyBrowser/1.2.3 } s₁    Accept-Language: en-US
Accept-Language: en-US              User-Agent: MyBrowser/1.2.3 } s₂
```

Figure 3.3. The sequence method illustrated using a simple HTTP request. Two different symbols s_1 and s_2 are encoded by the order of two selected header elements.

Two subpatterns of the sequence pattern exist. The *position pattern* includes hiding methods that signal information solely by the position of a particular PDU element within a number of other PDU elements. For instance, Rios et al. demonstrate that the position of a given DHCP option within the DHCP options list can be used to signal hidden information [5]. The second subpattern is the *number of elements* that encodes information by the number of elements in a (fragmented) PDU, for example, the number of options in an IPv4 header.

3.1.3 Add Redundancy to PDUs

The *add redundancy pattern* creates space within a given PDU or within a given PDU element. To create such space, redundant bits are artificially added to the transferred data and secret information is embedded into these redundant bits. A large number of hiding methods belong to this pattern. Wendzel et al. showed that this hiding pattern is one of the three that represent most of the known hiding methods [1]. For instance, Graf demonstrates the embedding of hidden data into new IPv6 destination options [7], Dyatlov and Castro extend HTTP headers with additional fields and extend values of existing header fields [4], and Trabelsi and Jawhar create space within IPv4 record route option fields [8].

3.1.4 Random Values in PDUs

Various network protocols comprise PDU fields that contain (pseudo-)random values, for example, the *Initial Sequence Number* (ISN) for freshly initiated TCP connections [9]. This allows one to embed encrypted covert content that exposes a similar distribution. Rowland demonstrates the placement of hidden data in the IPv4 Identifier and the TCP ISN fields [10], Rios et al. hide data in the DHCP *xid* field [5], and Lucena et al. utilize the SSH MAC field for the same purpose [11].

3.1.5 Value Modulation in PDUs

The *value modulation pattern* is one of the three most widely used patterns found in the study by Wendzel et al. [1]. The pattern includes hiding methods that select 1 of n possible values a header element can contain. For instance, a two bit field may be allowed to contain the values 00, 01, and 11, but not 10, and a hiding method signals hidden information using one of the three allowed values. Allowance for the selection of the n values can be due to different reasons, such as definitions of protocol specifications or given network environments. It is also imaginable that a field may contain all values that can be represented by its number of bits, but is not randomized, that is, not part of the add redundancy pattern. For instance, an IPv4 *time-to-live* (TTL) field can contain 256 possible values, but some of its values are unlikely to occur in a given network environment. Therefore, only n of the possible m values can be used to encode hidden information and for each packet sent, 1 of the n selected values can be chosen.

Girling demonstrated a covert channel that sends LAN frames to one of the n available Ethernet addresses in the local network [3]. Lucena et al. demonstrate the embedding of information in the IPv6 *hop limit* field [12]. A similar approach is available for the IPv4 TTL field [13].

Using one of the standardized IP protocol numbers, hidden information can be transferred by packets of a given protocol (e.g., an ICMP packet represents a ''0'' and a UDP packet represents a ''1'') [14]. Similarly, BACnet message types can be exploited by sending one of the allowed message types per packet [15].

Two subpatterns of the value modulation pattern exist. The *case pattern* comprises hiding methods that alter the case (upper or lower case) of plaintext header elements, such as in HTTP headers as demonstrated by Dyatlov and Castro [4] and visualized in Figure 3.4. The *LSB pattern* modifies only the least significant bit(s) of a header field. For instance, Handel and Sandford encode information using the even and odd times for the IPv4 timestamp option [16], and Giffin et al. modify the low-order bits of the TCP timestamp option [17].

3.1.6 Reserved/Unused Bits in PDUs

The last storage pattern comprises hiding methods that utilize reserved and unused fields in PDUs. Wendzel et al. concluded that no other pattern represents more hiding methods than the reserved/unused pattern—it represents 24 hiding methods [1]. This pattern can also be considered a trivial pattern, as unused and reserved fields are usually not interpreted by network stack implementations. However, changes of unused and reserved fields also result in a simple detection and prevention of the associated hiding methods.

Wolf demonstrated the embedding of hidden data into unused fields of data link layer frames [18], which was later extended by Handel and Sandford [16]. Lucena et al. list eight hiding methods for reserved and unused bits of the IPv6 protocol [12].

```
GET HTTP/1.1                    GET HTTP/1.1
Host: mywebsite.xyz             Host: mywebsite.xyz
USeR-AGEnt: MyBrowser/1.2.3     user-agENt: MyBrowser/1.2.3
s₁s₁s₂s₁  s₁s₁s₂s₂              s₂s₂s₂s₂  s₂s₁s₁s₂
```

Figure 3.4. Illustration of the case pattern. By only using one header line, multiple symbols per request can be transferred by modulating the case of letters.

3.2 HIDING INFORMATION IN THE TIMING OF PROTOCOL MESSAGES

In this section, we discuss techniques that encode hidden information in the timing of protocol messages, also referred to as timing covert channels. Protocol messages can belong to any layer of the network stack [19], for example, they can be link–layer frames, network–layer packets, or application protocol messages.

Timing channels are always noisy because of the timing inaccuracies at the covert sender and receiver. Other sources of noise are the loss of messages and network jitter, mainly caused by queue overflows and varying queuing delays in routers on the path between covert sender and receiver. The capacity of timing channels is often lower than that of noise-free storage channels, but they are potentially harder to detect and eliminate.

Some timing channels are protocol agnostic, which means they do not depend on features of the carrier network protocol, while others are protocol aware and depend on certain header fields and semantics of the carrier protocol (see Chapter 2). In this section, we first describe the protocol-agnostic techniques and then the protocol-aware techniques. Figure 3.5 shows the different types of timing methods we discuss (based on [1]).

3.2.1 Rate or Throughput of Network Traffic

Covert information can be encoded in a protocol-agnostic way by varying packet rates, which is equivalent to varying the throughput of network traffic. The covert sender varies its packet rate between two (binary channel) or multiple packet rates each time interval. The covert receiver measures the rate in each time interval and decodes the covert information. A binary channel can transmit one bit per time interval, whereas a multirate channel can transmit $\log_2 r$ bits per time interval, where r is the number of distinct packet rates or throughput values. Figure 3.6 shows an example of such a channel with binary encoding, that is, the covert sender encodes one bit per time interval.

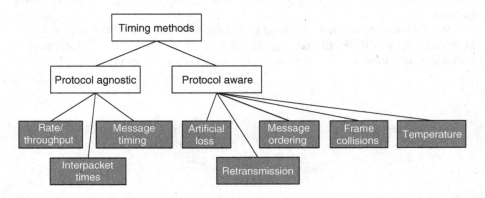

Figure 3.5. Taxonomy for network steganography timing methods.

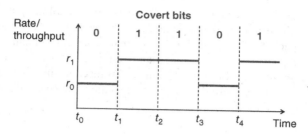

Figure 3.6. Example of using packet rate (throughput) to encode hidden communication. The covert sender encodes a zero bit as sending with rate r_0 and a one bit as sending with rate r_1. The covert receiver decodes the hidden messages based on the observed rates.

Padlipsky et al. outlined a timing channel where the sender either transmits or stays silent in each time interval [20] which was later implemented by Cabuk et al. [21]. This on/off timing channel is a special case of the binary channel where one rate is zero and the other is some chosen value.

Handel et al. proposed a throughput-based covert channel based on modulating the clear to send/ready to send (CTS/RTS) signals of serial port communication [16]. The CTS/RTS technique could be applied to other protocols utilizing CTS/RTS, such as IEEE 802.11 WLANs.

Luo et al. proposed to encode covert data into the length of TCP data bursts, where a data burst is the number of TCP segments sent between two TCP ACK arrivals [22]. The channel is more robust against noise, but has low capacity. Compared to other timing channels, its stealth is low, as covert channels behave very different from normal flows [22].

Packet rate steganography is noisy, since the sender's timing may not be accurate, packets may experience variable delays on the path (queuing delays), or packets may be lost on the path (e.g., dropped at congested routers). Girling suggested mitigating the noise problem by adjusting the packet rates used based on the noise observed [3].

A covert sender and a covert receiver need a mechanism for synchronization of the time intervals. One option is to have well-synchronized clocks and agree on the start times of the intervals. Another option is to solve the problem on the next-higher protocol layer, for example, Cabuk et al. [21] proposed to divide the covert data into small fixed-size frames, and synchronization is achieved through a special start sequence at the beginning of each frame. Cabuk et al. noted that their scheme still does not entirely solve the synchronization problem, so they proposed to use a phase lock loop (PLL) approach as improvement. The lack of adequate synchronization is another source of noise for packet rate channels.

Yao et al. studied the capacity of on/off packet rate channels based on packet rate distributions measured in real networks [23]. They showed that maximum information transmission rates of up to 180 bits/s could be achieved.

Li and He described a variant of the channel where a covert sender affects the performance of a switch to indirectly change the throughput of a packet flow from a third party to a covert receiver [24].

Mazurczyk and Szczypiorski propose a covert channel exploiting the IP protocol's fragmentation mechanism [25]. In this technique, the covert data are encoded depending on whether an IP packet is fragmented into an even number of fragments (zero bit) or an odd number of fragments (one bit). More than one bit per IP packet could be encoded by allowing a larger number of fragment counts. This mechanism is not strictly a packet rate channel, but it can be viewed as a related technique since it effectively encodes covert data as fragment rates.

3.2.2 Interpacket Times

Girling introduced in [3] a carrier protocol-independent packet timing channel that does not require synchronization of time intervals, because the covert sender encodes information in the timing of consecutive frames sent in a LAN. Berk et al. showed that the channel can also be used on the IP layer, where hidden information is encoded in the delays between IP packets, also called interpacket gaps [26]. The covert receiver decodes the information from the interarrival times of packets. Figure 3.7 shows an example of such a channel with binary encoding, that is, the covert sender encodes one bit per packet gap.

Berk et al. compared channels with two gap sizes (binary channels) and multiple gap sizes. They demonstrated how a covert sender can pick the optimal symbol distribution—the distribution that maximizes the channel capacity—in multisymbol channels given the channel characteristics based on the Arimoto–Blahut algorithm. However, the original approach was not very stealthy, as the packet gap distribution chosen by the covert sender could differ greatly from the packet gap distribution of normal traffic.

Gianvecchio et al. [27] and Sellke et al. [28] independently developed stealthier variants of the interpacket gap timing channel. They proposed to fit a model to the interpacket gap distribution of real traffic (without covert channel) and then use the model to generate covert channels with identical distribution. If the interpacket times of normal traffic are independent and identically distributed (i.i.d.), this channel is very hard to distinguish from the normal traffic with the detection techniques outlined by Gianvecchio and Wang [29].

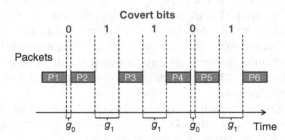

Figure 3.7. Example of using time gaps between packets to encode hidden communication. The covert sender encodes a zero bit as small gap g_0 and a one bit as large gap g_1. The covert receiver decodes the bits based on the gaps observed.

However, not all applications produce traffic with i.i.d interpacket times. In fact, some applications have correlated interpacket times [30]. To increase the stealth of the channel for applications with correlated interpacket times, Zander et al. [30] proposed to encode the covert information only in the least significant part of the interpacket gaps of existing traffic. This mimics the interpacket time distribution *and* correlations of normal traffic at the price of a reduced bit rate of the covert channel.

Liu et al. introduced a covert timing channel that encodes the covert data such that the normal distribution of interpacket times is closely approximated, and spreading techniques are used to provide robustness [31]. This channel is hard to detect with simple shape and regularity tests; however, these are known to be insufficient [29].

Wendzel et al. showed that interpacket gap timing channels can also be used with BACnet, a communications protocol for building automation and control networks [15].

Interpacket timing channels have been used by researchers to build a device that hooks into the connection between keyboard and computer, and exfiltrates all keystrokes by modulating the interpacket times of secure shell (SSH) network traffic send by the victim [32]. Other researchers have used the technique for embedding watermarks into packet flows (e.g., Houmansadr et al. [33]) in order to trace back traffic across proxies, anonymization networks, or stepping stones. These techniques are thoroughly discussed in Chapter 6.

While interpacket timing channels do *not* need synchronization of time intervals, they still suffer from the noise caused by the timing inaccuracies at covert sender and receiver, and the noise caused by variable packet delays or packet loss.

3.2.3 Message Sequence Timing

Wolf mentioned the possibility of constructing covert channels by modulating the use of protocol operations [18]. For example, a covert receiver can acknowledge each frame separately or wait until two frames have arrived before acknowledging the first. The general approach does not depend on features of the overt protocol, but concrete implementations depend on the actual protocol messages exchanged.

The timing of message acknowledgments (ACKs) can also be manipulated to construct a covert channel. This becomes an issue when a low-security entity wants to send data to a high-security entity *reliably*. Since reliable communication necessitates that the high-security entity sends ACKs for the data received, the timing of the ACKs can be manipulated to transmit covert data [34,35]. This channel is discussed in more detail in Section 8.4.4.

Eßer and Freiling implemented a web-based timing channel and analyzed its capacity [36]. In their scheme, a web server sends covert data to a client by delaying a response (a one bit) or by responding immediately (a zero bit). Li and Ephremides described timing channels in the Ad Hoc On-Demand Distance Vector (AODV) protocol [37]. The covert sender can modulate the times between successive AODV route requests, and the covert receiver can decode the information from the message timing.

Zou et al. proposed a technique for embedding covert channels into the File Transfer Protocol (FTP) [6]. Covert data are transmitted through varying the number of FTP

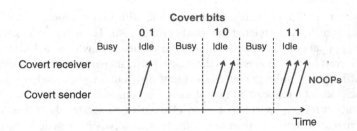

Figure 3.8. An FTP NOOP covert channel, an example of using message sequence timing for hidden communication. The integer value of the covert bits is encoded as the number of FTP NOOP commands sent during the idle periods when no data are transferred via FTP.

NOOPeration (NOOP) commands sent during idle periods; the number of NOOPs sent is equal to the integer value of the covert data encoded.

Figure 3.8 shows an example of how the FTP NOOP channel works. When the FTP session is idle, the covert sender sends a number of NOOP commands based on the covert bits to be transmitted. In our example, the covert sender sends only two bits per idle period, but in practice more bits could be send.

3.2.4 Artificial Message/Packet Loss

Servetto et al. demonstrated that channel erasures (message or packet loss) intentionally introduced at the sender can be used as a covert channel [38]. This is a protocol-aware technique since it requires per message/packet sequence numbers, so the receiver can detect the loss. Sequence numbers are present in many existing protocols. For example, TCP sequence numbers can be utilized, the IP ID field can be used if the sender produces consecutive IP IDs, and any application protocols with sequence numbers can also be used. Erasures are realized by artificially losing messages/packets at the sender.

Figure 3.9 shows an example of a channel that uses artificial packet losses to encode binary information. The covert sender encodes each one bit as packet loss and each zero bit as no packet loss.

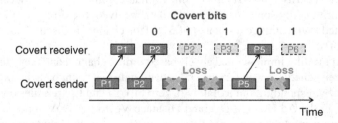

Figure 3.9. An example of using artificial packet loss to encode hidden communication. The covert sender encodes a zero bit as arrived packet and a one bit as artificially lost packet. The covert receiver decodes the information using the packet's sequence numbers.

Sadeghi et al. proposed a variation of the packet loss channel, where a man-in-the-middle (MITM) covert sender located between two virtual private network (VPN) sites drops selected packets exchanged between the VPN sites, in order to embed covert information into established connections between the VPN sites [39].

Packet loss channels do not suffer from timing noise, but lost or reordered packets may cause noise and reduce the channel capacity.

3.2.5 Artificial Retransmissions

Krätzer et al. proposed to transmit data covertly over IEEE 802.11 wireless LANs by duplicating frames [40]. The covert sender duplicates frames for selected connections to send a one bit, and duplicates frames for other selected connections to send a zero bit. Covert sender and receiver must agree beforehand on a matrix that defines the "connections" (source and destination MAC addresses) used for encoding and a key matrix that defines for each connection whether duplications encode a logical zero or one.

However, there are many other possibilities for encoding hidden data in retransmissions of frames, packets, or messages. For example, selected DNS requests can be sent once or twice to encode a hidden bit per request [1].

There are also hybrid covert channels that utilize packet loss and retransmissions. Section 3.3 describes a channel in which the covert receiver does not acknowledge a successfully received packet. The covert sender then retransmits the packet, but the payload of the retransmitted data contains a steganogram instead of the user data.

3.2.6 Manipulated Message Ordering

Kundur and Ahsan described a covert channel implemented through packet reordering [41]. Because a set of n packets (or messages) can be arranged in any $n!$ ways, a maximum of $\log_2 n!$ bits can be transmitted. This approach is protocol aware, since it requires per packet (or message) sequence numbers that the covert receiver can use to determine the original packet order.

One way this technique could be implemented is to reorder TCP packets, as each TCP packet has a sequence number. However, the technique could also be implemented with IP packets if the IP ID field has unique values, or with any application-layer protocol that has sequence numbers. Another possibility is to modify the order of IPSec Authentication Header (AH) packets or IPSec Encapsulated Security Payload (ESP) packets [42].

Figure 3.10 shows a simple example of encoding hidden information in packet reordering. Here, the covert sender sends a packet at the correct position to encode a zero bit, and a packet in an incorrect position to encode a one bit.

Chakinala et al. proposed a formal model for transmitting information via packet-reordering [43]. They developed several channel and jamming models. Using a game-theoretic approach, they modeled the channel as game between covert sender and receiver on one side and the jammer on the other, and proved the existence of a Nash equilibrium for the mutual information rate.

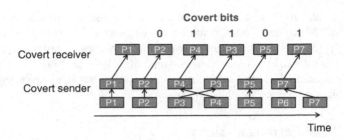

Figure 3.10. An example of (re)ordering packets to encode hidden communication. A packet in a correct position encodes a zero bit, while a packet in an incorrect position encodes a one bit.

Luo et al. developed a method that encodes covert information on the order of N packets across X flows [44]. Depending on whether single packets and/or flows can be distinguished from each other, there are various ways of encoding the covert data. If one thinks of flows as urns and packets as balls, then these encodings are directly related to the counting problem of drawing N balls from X urns.

Atawy and Al-Shaer developed another covert channel based on packet reordering [45]. They used fake IP traffic with sequence numbers embedded in the payload. However, since the payload does not look like normal traffic, any closer inspection would reveal the covert channel.

Reordering channels—like loss channels—suffer from noise caused by message/packet loss or reordering.

3.2.7 Collision and Timing of Frames

Handel and Sandford proposed exploiting the Ethernet Carrier Sense Multiple Access Collision Detection (CSMA/CD) mechanism [16]. If frames collide in CSMA/CD, a jamming signal is issued and the senders back off a random amount of time. The covert sender deliberately jams packets of another user. Then, it uses a back off delay of either zero or the maximum value. Therefore, all frames sent by the covert sender will either lead or lag packets sent by the other user, essentially creating a one bit per frame collision covert channel. The receiver can recover the information by detecting the collisions and analyzing the order of frame arrivals after the collisions.

The technique was initially proposed with the CSMA/CD approach of Ethernet in mind, but it can be adapted to other carrier protocols with CSMA mechanisms, such as IEEE 802.11 wireless LAN protocols.

Figure 3.11 shows an example of the frame jamming covert channel. The covert sender encodes a zero bit as back off with zero delay d_0 and a one bit as back off with delay d_1.

To improve performance of shared medium access protocols, splitting algorithms are used to divide the set of collided senders into smaller subsets. Then, these subsets retransmit in order. Dogu and Ephremides designed a covert channel using the first come first serve (FCFS) splitting algorithm [46]. The covert information is conveyed

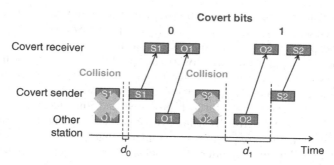

Figure 3.11. An example of using frame jamming for hidden communication. To send a zero or one bit, the covert sender retransmits with delay d_0 or delay d_1, respectively, after a previous frame collision.

in the number of collisions observed in a collision resolution period. The covert sender controls this number by generating dummy packets that cause additional collisions. The covert receiver passively monitors the channel and keeps track of the collision resolution procedure to extract the covert information.

Li and Ephremides' covert transmission scheme uses the covert sender's splitting decisions (which subsets it joins) as a carrier of covert data [47]. The covert receiver passively tracks the collision resolution procedure. When it detects a successful transmission from the covert sender, it can infer past splitting decisions, which is the encoded covert data.

3.2.8 Temperature-Based Covert Channels

Murdoch proposed a protocol-aware channel where the covert sender encodes bits as varying service request rate, and the covert receiver decodes the original signal from estimating the changes in clock difference (also called *clock skew*) [48]. The channel requires an intermediate host that receives and sends packets to both covert sender and receiver. The channel exploits the fact that a host's CPU temperature depends on the number of service requests per time unit it processes, and the skew of a host's system clock depends on the temperature.

On the receiving side, the channel requires that the intermediate supports a protocol with timestamps from which the clock skew can be estimated. Possible options are TCP, if TCP timestamps are enabled (these are now enabled by default on many OS), ICMP echo messages, HTTP, which since version 1.1 has timestamps enabled by default, or other application protocols with timestamps.

The covert sender either sends requests to the intermediary (one bit) or stays silent (zero bit), thus changing the temperature and thereby indirectly the clock skew on the intermediate host. The covert receiver obtains clock samples from responses of the intermediate host (e.g., HTTP timestamps in response to the covert receiver's GET requests) and uses these to estimate the clock skew on the intermediate host. From the estimated clock skew, the covert receiver can decode the covert bits. Murdoch [48] demonstrated that information can be transmitted successfully over this channel, but Zander et al. showed that its capacity is low (less than 20 bits/h) [49].

The low capacity is due to various factors, such as noisy CPU–load temperature coupling, sender throughput limiting networks, or variable packet delays on the path from an intermediary node to the covert receiver. However, with low-resolution timestamps, such as the timestamps in the HTTP protocol, the main source of noise is the timestamp quantization error. Zander and Murdoch proposed and implemented a technique to minimize this quantization error by using synchronized sampling; the covert receiver synchronizes its timestamp requests with the intermediate's clock ticks, attempting to obtain timestamps immediately after clock ticks, where the quantization error is the smallest [50].

The channel was originally developed to reveal hidden services (e.g., web servers) that provide mutual anonymity[1] inside the Tor network [51]. The attacker identifies the hidden service by correlating variable clock skew patterns measured over time for several candidates with the request rate pattern sent to the hidden service over the anonymization network. This only requires a small number of bits to be transmitted. However, the channel could be used for other purposes, as long as the covert messages are short (e.g., exfiltrated passwords).

Figure 3.12 depicts the temperature-based covert channel in the scenario where covert sender and receiver are separated from the intermediate host by a network. The covert sender and receiver can be controlled by the same person (e.g., attacking Tor hidden services), or can be different persons (e.g., general covert communications).

3.2.9 Indirect Timing Channels

Unlike direct timing channels, with indirect timing channels there is no direct exchange of timing information between covert senders and receivers, which improves the stealth. However, these channels are harder to construct, and only very few proposals exist.

Hintz described an indirect timing channel using a public server as intermediate host [52]. The covert sender sends a large number of requests to the server or stays silent in each time interval, equivalent to one bit per time interval. The covert receiver periodically probes the server and measures the response time to recover the covert information.

Another example of an indirect timing channel is the temperature-based timing channel described in Section 3.2.8.

3.2.10 Covert Sender Location

As explained in Chapter 2, the covert sender can either generate cover traffic with an embedded covert channel that reaches the covert receiver, or act as man-in-the-middle by embedding the covert channel into existing overt traffic between unwitting

---

[1] The identity of the service operator is not revealed to the user and the identity of the user is not revealed to the service operator.

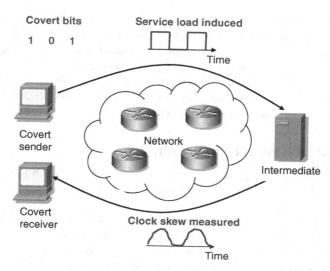

Figure 3.12. A temperature-based covert channel. The covert sender encodes information by changing the CPU load on the intermediate host through changing the service request rate. The CPU load changes affect the temperature, which in turn affects the clock skew on the intermediate host. The covert receiver measures the clock-skew change over time to reconstruct the original load pattern and thereby decode the covert bits. (Reproduced from [63] with permission of IEEE.)

senders/receivers that reaches the covert receiver. MITM channels are stealthier, but in many cases also have lower capacity.

All timing methods mentioned in the previous sections can be realized as non-MITM channels. Many of the rate, interpacket gap, message timing, and loss or reordering techniques can also be used to create MITM channels. However, with MITM channels, there are tighter limits as to how much the covert sender can manipulate the overt traffic before the overt senders or receivers notice something. The frame collision channel and the indirect channel methods described can only be realized as non-MITM channels, since the covert sender *must* generate some traffic.

3.3 HYBRID METHODS

The complexity of network steganography methods has been growing continuously over the time mainly due to increasingly sophisticated network protocols and services in communication networks. Hybrid solutions, that is, the methods that modify protocol fields and protocol messages timing, are a natural evolution from the previously mentioned two main subgroups of network steganography techniques.

Typically, a nonhybrid network steganography method utilizes a single subcarrier for covert communication purposes. Let us consider, for example, a storage method that is based on modification of the IPv4 type of service (ToS) field. In this case, a subcarrier is this particular field in which the secret data are embedded. This is similar for many other methods, for example, those that are based on reordering of the packets, or for the LSB method applied to digital media signals. In contrast, hybrid techniques make use of both timing and storage methods, therefore, they require two subcarriers.

In the following sections, we will review two examples of hybrid network steganography that follow the above-mentioned pattern: LACK (lost audio steganography) [53] and RSTEG (retransmission steganography) [54].

3.3.1 Lost Audio Packets Steganography

Lost Audio Packets steganography was originally proposed in 2008 [53] and studied further later on, [55–57]. Currently, it is considered as one of the state-of-the-art VoIP steganographic techniques [58].

LACK is an IP telephony steganographic method, which modifies both RTP packets from the voice stream and their time dependencies. This method takes advantage of the fact that in typical multimedia communication protocols, like RTP, excessively delayed packets are not used for the reconstruction of transmitted data at the receiver, that is, the packets are considered useless and discarded.

The detailed overview of LACK is presented in Figure 3.13. At the transmitter (Alice), one RTP packet is selected from the voice stream and its payload is substituted

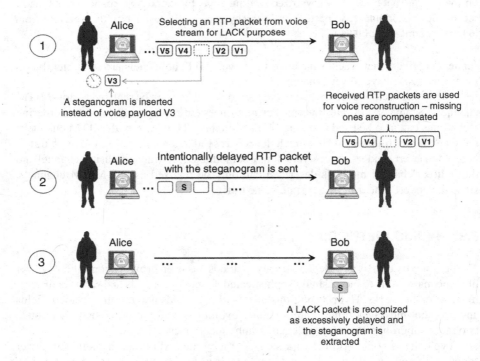

Figure 3.13. The idea of LACK. (Reproduced from [34] with permission of Wiley.)

with bits of the secret message (1). Then, the selected audio packet is intentionally delayed prior to its transmission (2). Whenever an excessively delayed packet reaches a receiver unaware of the steganographic procedure, it is discarded—it is treated as if it has never arrived. However, if the receiver (Bob) is aware of the hidden communication, then, instead of dropping the received RTP packet, it extracts the payload (3). The payload of the intentionally delayed packets is the sole vector used to transmit secret information to receivers aware of the procedure, and no additional packets are generated.

LACK is a hybrid approach because it utilizes two different subcarriers, and, as a result, two information hiding techniques: timing and storage (Figure 3.14 (1)). The timing method is used for the selection of the suitable RTP packet for LACK purposes and the storage method is responsible for embedding secret data into the payload of the chosen RTP packet. This is different than for typical network steganography methods where only one subcarrier is utilized, similar to the ToS-based information hiding case illustrated in Figure 3.14 (2).

A solution that can be treated as a variation of the LACK method was also proposed in 2012 by Arackaparambil et al. [59]. In this paper, authors described a simple VoIP steganography method in which the payloads of selected RTP packets are replaced with a steganogram and the RTP header's sequence number, and/or timestamp fields are intentionally changed to make them appear as if they were excessively delayed by the network. The concept of LACK was further extended by Hamdaqa and Tahvildari [58] by providing a reliability and fault tolerance mechanism based on a modified (k, n)

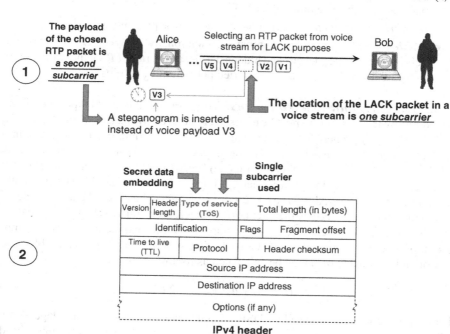

Figure 3.14. LACK as an example of a hybrid method.

threshold based on Lagrange interpolation, and results demonstrated that the complexity of steganalysis is increased. The ''cost'' for the extra reliability is a loss of some fraction of the steganographic bandwidth.

LACK is a TCP/IP stack application layer steganography technique and is fairly easy to implement [57]. This may be attributed to the fact that RTP is usually integrated into telephone endpoints (softphones). Therefore, access, generation, and modification of RTP packets is easier than for lower layer protocols like IP or UDP.

The general principle in LACK is that the more the hidden information inserted into the voice stream, the greater the chance that it will be detected, that is, by scanning the data flow or applying some other steganalysis (detection) method. Hence, the more the audio packets used to send covert data, the greater the deterioration of the quality of the IP telephony connection. This, in turn, results in a greater steganographic cost. Therefore, the procedure of inserting the hidden data must be carefully chosen and controlled in order to minimize the chance of the detection of the inserted data and to avoid excessive deterioration of the quality of service (QoS). That is why certain trade-offs between the achievable steganographic bandwidth, call quality deterioration and resistance to detection, are indispensable. The performance of LACK depends on many factors that can be divided into three following groups:

- *Endpoint-related factors*: The type of voice codec used (in particular, its resistance to packet losses and the default voice quality), size of the RTP packet payload, and the size of the jitter buffer.
- *Network-related factors*: Packet delay, packet loss probability, and jitter.
- *LACK-related factors*: The number of intentionally delayed RTP packets, the delay of the LACK packets, and hidden data insertion rate (IR), which correspond to the number of secret data bits carried per unit of time (bit/s).

3.3.1.1 *Endpoint-Related Factors.* To guarantee that an RTP packet will be deemed lost by the receiver, it must be excessively delayed by the LACK procedure. To set this delay $d_L(t)$ properly, the size of the receiver's jitter buffer must be taken into account. A jitter buffer is used to alleviate the jitter effect, that is, the variations in packet arrival times caused by queuing, contention, and serialization in the network. The size of the buffer is implementation dependent. It may be fixed or adaptive, and is usually between 60 and 120 ms. An RTP packet will be recognized as lost whenever its delay exceeds the delay introduced by the jitter buffer. LACK users must exchange information about the sizes of their jitter buffers prior to starting the hidden communication. To limit the risk of disclosure, the delay chosen by LACK should be as low as possible. The delay of an RTP packet (d_T) may be calculated as follows

$$d_T(t) = d_D + d_K + d_E + d_L(t), \tag{3.1}$$

where d_D is the delay introduced by DSP (digital signal processor), which depends on the type of the codec and typically ranges from 2 to 20 ms, d_K is the delay introduced by voice coding (typically under 10 ms), d_E is the delay caused by encapsulation (from 20 to 30 ms), and $d_L(t)$ is the intentional delay of an RTP packet introduced by LACK.

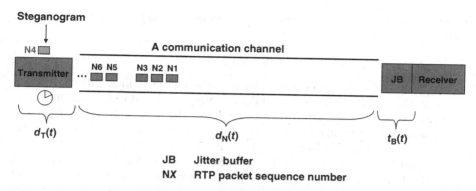

JB   Jitter buffer
NX   RTP packet sequence number

Figure 3.15. Components of the LACK delay. (Reproduced from [34] with permission of Wiley.)

As mentioned above, the value of the intentional delay $d_L(t)$, introduced by LACK, must be carefully chosen. Together with $d_N(t)$, introduced by the network, it must exceed the size of the jitter buffer (Figure 3.15), that is,

$$d_T(t) + d_N(t) > t_B(t), \tag{3.2}$$

where $d_N(t)$ is the delay introduced by the network and $t_B(t)$ is the size of the jitter buffer (denotes typically the time that a packet spends in the jitter buffer and is expressed in seconds).

The jitter buffer can be of a fixed or adaptive size. If the jitter buffer has a fixed size during the call, and information about delay caused by the network is not taken into account, then the delay at the transmitter side should be

$$d_T \geq t_B \tag{3.3}$$

and

$$d_L \geq t_B - d_D - d_K - d_E. \tag{3.4}$$

Consequently, if the buffer is of an adaptive size and it reacts to the current delay introduced by the network ($d_N(t)$), then the delay at the transmitter output is

$$d_T(t) \geq t_B(t) - d_N(t). \tag{3.5}$$

Thus,

$$d_L(t) \geq t_B(t) - d_D - d_K - d_E - d_N(t). \tag{3.6}$$

If the current value of $d_N(t)$ is not known at the transmitter, then one can utilize the average value of the delay calculated over a certain time period. If the adaptive jitter buffer is used at the receiver, and the information regarding its size is not passed to the transmitter during the call, then the relation $d_T(t) \geq t_B(t)$ should be fulfilled. This is to ensure that intentionally delayed RTP packets will not be used for voice reconstruction.

Due to the fact that delays d_D, d_K and d_E are constant, ensuring $d_T(t) \geq t_B(t)$ is possible when

$$d_L(t) \geq t_B^* - d_D - d_K - d_E, \qquad (3.7)$$

where t_B^* denotes the maximum, admissible size of the adaptive jitter buffer.

Under the considered conditions, if the receiver is equipped with an adaptive jitter buffer and it is possible to advertise its size during the call, then its initial size can be communicated during the signaling phase of the call. This imposes that the delay at the transmitter output $d_T(0)$ is set equal to the maximum possible size of the jitter buffer: $d_T(0) \geq t_B^*$. It can be further decreased by means of reducing $d_L(t)$, which is possible if appropriate information about the variations in size of the jitter buffer reaches the transmitter during the call. When an adaptive jitter buffer is employed, and the transmitter is informed of the current network delay $d_N(t)$, then

$$d_L(t) \geq t_B(t) - d_D - d_K - d_E - d_N(t). \qquad (3.8)$$

The other factor that influences LACK is the VoIP codec used for the conversation. The greater a codec's resistance to packet losses is, the more favorable it is for LACK purposes. The admissible level of packet losses for different voice codecs usually ranges from 1 to 5%. For example, according to Na and Yoo [60], the maximum loss tolerance equals 1% for G.723.1, 2% for G.729A, and 3% for G.711 codecs. The usage of mechanisms that deal with lost packets at the receiver, for example, packet loss concealment (PLC), results in an increase in the acceptable level of packet losses, for example, for G.711 the shift is from 3 to 5%. Thus, as mentioned, the greater the codec's resistance to packet losses, the greater the potential steganographic bandwidth. Therefore, the quantity of covert data liable for insertion by LACK procedure and, consequentially, the additional induced packet losses depend on the acceptable level of the cumulative packet loss. It is also worth noting that the use of a silence suppression mechanism in the transmitting endpoint can further decrease the available steganographic bandwidth.

3.3.1.2 Network-Related Factors. Let us assume that, at a given moment of the call t, an RTP packet is chosen from the voice packets stream for LACK purposes with probability $p_L(t)$ and the network packet loss probability is $p_N(t)$. If p_T denotes the maximum permitted probability of RTP packet losses, then, assuming the independence of the network-related losses from LACK-induced losses, we get

$$p_T \leq 1 - (1 - p_N(t))(1 - p_L(t)), \qquad (3.9)$$

and, in consequence,

$$p_L(t) \geq \frac{p_T - p_N(t)}{1 - p_N(t)}, \qquad (3.10)$$

which describes the admissible level of RTP packet losses introduced by LACK. Exemplary relationships between probabilities $p_L(t)$, $p_N(t)$, and p_T are illustrated in Figure 3.16. To ensure a high steganographic bandwidth and the undetectability of LACK, it is necessary to monitor network conditions while the call lasts. In particular,

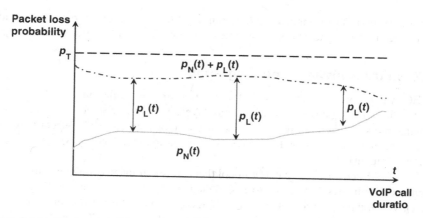

Figure 3.16. The impact of LACK on the total packet loss probability. (Reproduced from [57] with permission of Wiley.)

packet losses, delay, and jitter introduced by the network must be observed. They have influence on the range of permitted values of the delay and packet losses introduced by LACK without the degradation of the perceived quality of the conversation. Due to the fact that LACK exploits legitimate RTP traffic, an increase in the overall packets losses is triggered. Thus, the number of lost packets used for steganographic purposes must be controlled and dynamically adapted.

Information concerning current network conditions can be provided to the transmitter, among others, with the aid of SR (sender report), RR (receiver report) [61], or XR (extended report) [62], that are defined in the RTCP protocol. The lack of monitoring of network parameters (e.g., packet delays and losses) during a call does not hinder the possibility of determining their values, which can be achieved with the aid of historical, statistical data related to the network performance. However, it should be noted that RTP packet losses introduced by the network can decrease LACK's steganographic bandwidth, if the lost packet(s) contains secret data.

3.3.1.3 LACK-Related Factors.

In previous sections, we mentioned two important parameters that require setting for proper LACK functioning. These are the probability that a certain RTP packet is chosen for LACK purposes $(p_L(t))$, and the delay $d_L(t)$, which is preset to guarantee that an audio packet will be recognized as lost at the receiver. As mentioned earlier, a key factor influencing LACK steganographic bandwidth and its resistance to steganalysis is the hidden data insertion rate $IR(t)$. In general, the greater, the $IR(t)$, the greater the achievable steganographic bandwidth. This couples with the degradation of voice quality and easier steganalysis. The limits imposed on the maximum insertion rate depend on the targeted acceptable call quality, network conditions, the size of the steganogram, and also the duration of the call. The correct determination of $IR(t)$ facilitates efficient control of RTP packet losses and delays introduced by LACK, without excessively deteriorating the call quality and risking detection. The methods for determining $IR(t)$ based on current conversation quality, the size of the steganogram, and the duration of the call are described in [55,56].

3.3.2 Retransmission Steganography

RSTEG was proposed in 2011 in [54] and was further extended in [63]. RSTEG can be used for all protocols at different OSI layers that utilize generic retransmission mechanism (as depicted in Figure 3.17). It may also be applied to TCP-specific retransmission mechanisms, such as FR/R (fast retransmit and recovery) or SACK (selective acknowledgment).

In a simplified case, a typical protocol that uses a retransmission mechanism based on timeouts obligates a receiver to acknowledge each received packet (Figure 3.17, case 1). When the packet is not successfully received, no acknowledgment is sent after the timeout expires, and so the packet is retransmitted (Figure 3.17, case 2).

RSTEG uses a retransmission mechanism to reliably exchange secret data (shown in Figure 3.18). Both a sender and a receiver are aware of the steganographic procedure. At some point during the connection after successfully receiving a packet, the receiver intentionally does not issue an acknowledgment message. In a normal situation, a sender is obligated to retransmit the lost packet when the time frame within which the packet acknowledgment should have been received expires. In the context of RSTEG, a sender replaces the original payload with secret data instead of sending the same packet again. When the retransmitted packet reaches the receiver, it can extract hidden information (Figure 3.18, case 3).

Note that the hidden data sender and receiver may not be simultaneously the sender and receiver of the overt communication; in some scenarios, both can be located at network intermediate nodes. In this case, it is harder to uncover the steganographic

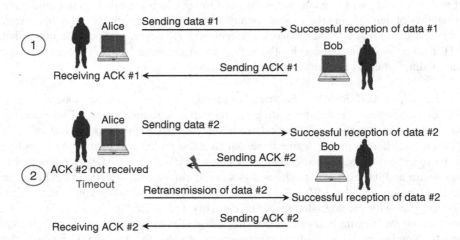

Figure 3.17. Generic retransmission mechanism based on timeouts. (Reproduced from [63] with permission of Springer.)

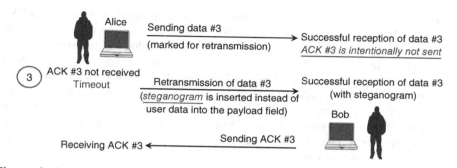

Figure 3.18. The concept of retransmission steganography. (Reproduced from [63] with permission of Springer.)

communication as the typical location of the node used for steganalysis is near the sender or receiver of the overt transmission.

Like LACK, RSTEG is a hybrid approach because it utilizes a timing method to indicate the packet that will be retransmitted and a storage one for the embedding of secret data. Thus, two subcarriers are utilized: one is a retransmission mechanism and the other is the selected packet's payload.

The performance of RSTEG depends on many factors, such as the details of the communication procedure (in particular, the size of the packet payload, the rate at which segments are generated, etc.). No real-world steganographic method is perfect; whatever the method, the hidden information can be potentially discovered. In general, the more the hidden information inserted into the data stream, the greater the chance that it will be detected, for example, by scanning the data flow or by some other steganalysis methods. However, some steps can be taken for RSTEG in order to avoid easy detection. For instance, in [64], Zhai et al. proposed improvements to the original RSTEG design by ensuring that the checksum field of the original and the retransmission segments are the same, thus enhancing its undetectability.

Applying RSTEG to TCP is the natural choice for IP networks, as a vast amount of Internet traffic (about 80–90%) is based on this protocol. For TCP, the following retransmission mechanisms are defined:

- RTO (retransmission timeouts) in which segment loss detection is based on RTO timer expiration. Results from [65] show that 60–88% of all retransmissions on the Internet were caused by RTO mechanism. In RTO, a segment is considered lost if the receiver does not receive an acknowledgment segment (ACK) after the specified period of time, after which it is retransmitted. The RTO timer value varies in TCP implementations across different operating systems, and it depends mainly on the RTT (round-trip time) and its variation. When the RTO timer is set to a value too low, it may cause too many spurious retransmissions; otherwise, the sender will be waiting too long to retransmit a lost segment, which may cause throughput decrease.

- FR/R (fast retransmit/recovery) is based on detecting duplicate ACKs (i.e., ACKs with the same acknowledgment number). A receiver acknowledges all segments delivered in order. When segments arrive out of order, the receiver must not increase the acknowledgment number so as to avoid data gaps, but instead must send ACKs with unchanged acknowledgment number values, which are called duplicate ACKs (dupACKs). Usually, a segment is considered lost after the receipt of three duplicate ACKs. Issuing duplicate ACKs by the receiver is often a result of out-of-order segment delivery. If the number of duplicate ACKs that triggers retransmission is too small, it can cause too many retransmissions and can degrade network performance.

- SACK (selective acknowledgment) is based on fast retransmit/recovery. It uses an extended ACK option that contains block edges to deduce which received blocks of data are noncontiguous. When retransmission is triggered, only missing segments are retransmitted. This feature of SACK decreases network load.

The intentional retransmissions caused by RSTEG should be kept at a reasonable level to avoid detection. To achieve this goal, it is necessary to determine the average number of natural retransmissions in TCP-based Internet traffic as well as to know how intentional retransmissions affect the network retransmission rate. Usually network retransmissions are caused by network congestion, excessive delays, or reordering of packets [65], and their number is estimated to account for up to 7% of all Internet traffic [65,66].

RSTEG can be applied to all retransmission mechanisms presented above. It requires modification to both a sender and a receiver. A sender should control the insertion procedure and decide when a receiver should invoke a retransmission. The sender is also responsible to keep the number of retransmissions at a nonsuspicious level. The receiver's role is to detect when the sender indicates that intentional retransmission should be triggered. Then, when the retransmitted segment arrives, the receiver should be able to extract the secret data.

The sender must be able to mark segments selected for a hidden communication (i.e., retransmission request segments) so that the receiver can determine for which segments to invoke retransmissions and which will contain secret data. However, marked TCP segment should not differ from those sent during a connection. The following procedure for marking sender segments is proposed. Let us assume that the sender and receiver share a secret *Steg-Key* (SK). For each fragment chosen for a steganographic communication, the following hash function *(H)* is used to calculate the *identifying sequence (IS)*:

$$IS = H(SK||SequenceNumber||CB), \hspace{2cm} (3.11)$$

H can be any secure hash function. *Sequence number* denotes the value from the chosen TCP header field, || is the bits concatenation function, and CB is a control bit that allows the receiver to distinguish a retransmission request segment from a segment with secret data embedded. For every TCP segment used for hidden communications, the resulting IS will have a different value due to the monotonically increasing TCP sequence number. All IS bits (or only selected ones) are distributed by the sender across a segment's payload field in a predefined manner. The receiver must analyze each incoming segment; based on SK and values from the TCP header, the receiver calculates two values of IS, namely, one with CB = 1 and one with CB = 0. Then the receiver checks if and which IS is present inside the received segment.

Problems may arise when the segment that informs the receiver of a necessity to invoke an intentional retransmission (which contains user data together with the IS) is lost due to network conditions. In that case, a normal retransmission is triggered, and the receiver is not aware that the segment with hidden data will be sent. However, in this case, the sender believes that the retransmission was invoked intentionally by the receiver, and so he/she issues the segment with steganogram and the IS. In this scenario, user data will be lost, and the cover connection may be disturbed.

In order to address the situation in which the receiver reads a segment with an unexpected steganogram, the receiver should not acknowledge the reception of this segment until he/she receives the segment with user data. When the ACK is not sent to the sender, another retransmission is invoked. The sender is aware of the data delivery failure, but he/she does not know which segment to retransmit, so he/she first issues a segment with user data. If the delivery confirmation is still missing, then the segment with the secret data is sent. The situation continues until the sender receives the correct ACK. This mechanism for compensating network losses is illustrated in Figure 3.19.

This mechanism introduces substantial delay, but in practice it is not used frequently. For example, consider the scenario in which 0.5% intentional retransmissions are invoked. If 5% of them are lost, it means that the above-described mechanism will take place only for 0.025% of steganographic segments, thus it will occur rarely.

RSTEG may be applied to the retransmission mechanisms presented previously as follows:

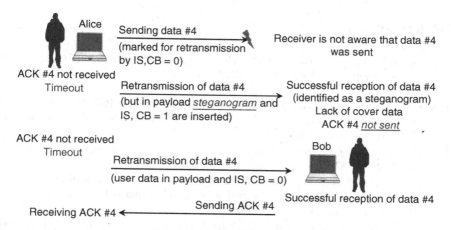

Figure 3.19. An example of the RTO-based RSTEG segment recovery. (Reproduced from [38] with permission of Springer.)

- *RTO-based RSTEG:* The sender marks a segment selected for hidden communication by distributing the IS across its payload. After successful segment delivery, the receiver does not issue an ACK message. When the RTO timer expires, the sender sends a steganogram inside the retransmitted segment's payload. The receiver extracts the steganogram and sends the appropriate acknowledgment.
- *FR/R-based RSTEG:* The sender marks the segment selected for hidden communication by distributing the IS across its payload. After successful segment delivery, the receiver starts to issue duplicate ACKs to trigger retransmission. When the ACK counter at the sender side exceeds the specified value, the segment is retransmitted. The payload of the retransmitted segment contains secret data. The receiver extracts the secret data and sends an appropriate acknowledgment.
- *SACK-based RSTEG:* The scenario is exactly the same as FR/R, but in the case of SACK, it is possible that many segments are retransmitted because of the potential noncontiguous data delivery.

3.4 SUMMARY

This chapter introduced various information hiding methods based on the modification of protocol fields or the timing of protocol messages, as well as hybrid approaches that combine the modification of protocol fields and protocol message timing to signal hidden information. We discussed the techniques in relation to the taxonomy introduced in Chapter 2. We made no attempts to assess which methods are the most suitable today with respect to capacity or undetectability. Such an assessment is not obvious, as capacity or undetectability depends on the circumstances in which a channel is used (e.g., amount of covert data, available cover traffic), and remains future work.

REFERENCES

1. S. Wendzel, S. Zander, B. Fechner, and C. Herdin. A pattern-based survey and categorization of network covert channel techniques. *ACM Computing Surveys* 47(3), pp. 50:1–26, ACM, 2015.
2. S. J. Murdoch and S. Lewis. Embedding covert channels into TCP/IP. In *Proceedings of the Information Hiding Conference 2005*, Vol 3727 of *Lecture Notes in Computer Science*, pp. 247–261. Springer, 2005.
3. C. G. Girling. Covert channels in LAN's. *IEEE Transactions on Software Engineering*, 13(2):292–296, 1987.
4. A. Dyatlov and S. Castro. Exploitation of data streams authorized by a network access control system for arbitrary data transfers: tunneling and covert channels over the http protocol. Technical Report, Gray-World.net, 2005.
5. R. Rios, J. Onieva, and J. Lopez. HIDE_DHCP: covert communications through network configuration messages. In *Proceedings of the IFIP TC 11 27th International Information Security Conference*. Springer, 2012.

6. X. Zou, Q. Li, S. Sun, and X. Niu. The research on information hiding based on command sequence of FTP protocol. In *Proceedings of the 9th International Conference on Knowledge-Based Intelligent Information and Engineering Systems*, pp. 1079–1085, September 2005.

7. T. Graf. Messaging over ipv6 destination options, 2003. http://gray-world.net/papers/messip6.txt, December 2013.

8. Z. Trabelsi and I. Jawhar. Covert file transfer protocol based on the IP record route option. *Journal of Information Assurance and Security*, 5(1):64–73, 2010.

9. S. Bellovin. RFC 1948: defending against sequence number attacks. https://www.ietf.org/rfc/rfc1948.txt, February 2015.

10. C. H. Rowland. Covert channels in the TCP/IP protocol suite. *First Monday*, 2(5), May 1997. http://firstmonday.org/htbin/cgiwrap/bin/ojs/index.php/fm/article/-view/528/449, February 2015.

11. N. Lucena, J. Pease, P. Yadollahpour, and S. J. Chapin. Syntax and semantics-preserving application-layer protocol steganography. In *Proceedings of the 6th Information Hiding Workshop*, May 2004.

12. N. Lucena, G. Lewandowski, and S. Chapin. Covert channels in IPv6. In *Proceedings of the 5th International Workshop on Privacy Enhancing Technologies (PET '05)*, Vol. 3856 of *Lecture Notes in Computer Science*, pp. 147–166. Springer, 2006.

13. S. Zander, G. Armitage, and P. Branch. Covert channels in the IP time to live field. In *Australian Telecommunication Networks and Applications Conference (ATNAC '06)*, pp. 298–302, 2006.

14. S. Wendzel. Protocol channels as a new design alternative of covert channels. Technical Report, CORR, Vol. abs/0809.1949, Kempten University of Applied Sciences, September 2008. http://arxiv.org/abs/0809.1949.

15. S. Wendzel, B. Kahler, and T. Rist. Covert channels and their prevention in building automation protocols: a prototype exemplified using BACnet. In *Proceedings of the 2nd Workshop on Security of Systems and Software Resiliency*, pp. 731–736. IEEE, 2012.

16. T. Handel and M. Sandford. Hiding data in the OSI network model. In *Proceedings of the First International Workshop on Information Hiding*, pp. 23–38, 1996.

17. J. Giffin, R. Greenstadt, P. Litwack, and R. Tibbetts. Covert messaging through TCP timestamps. In *Proceedings of the 2nd International Conference on Privacy Enhancing Technologies*, pp. 194–208. Springer, 2003.

18. M. Wolf. Covert channels in LAN protocols. In *Proceedings of the Workshop on Local Area Network Security (LANSEC)*, pp. 91–101, 1989.

19. I. standard 7498-1. Information technology: open systems interconnection—basic reference model, 1994.

20. M. A. Padlipsky, D. W. Snow, and P. A. Karger. Limitations of end-to-end encryption in secure computer networks. Technical Report, ESD-TR-78-158, Mitre Corporation, August 1978. http://stinet.dtic.mil/cgi-bin/GetTRDoc?AD=A059221&Location=U2&doc=GetTRDoc.pdf.

21. S. Cabuk, C. E. Brodley, and C. Shields. IP covert timing channels: design and detection. In *Proceedings of the 11th ACM Conference on Computer and Communications Security (CCS)*, pp. 178–187, October 25–29, 2004.

22. X. Luo, E. W. Chan, and R. K. Chang. TCP covert timing channels: design and detection. In *Proceedings of the IEEE/IFIP International Conference on Dependable Systems and Networks (DSN)*, June 2008.

23. L. Yao, X. Zi, L. Pan, and J. Li. A study of on/off timing channel based on packet delay distribution. *Computers & Security*, 28(8):785–794, 2009.

24. W. Li and G. He. Towards a protocol for autonomic covert communication. In *Proceedings of the 8th International Conference on Autonomic and Trusted Computing*, pp. 106–117, 2011.

25. W. Mazurczyk and K. Szczypiorski. Evaluation of steganographic methods for oversized IP packets. *Telecommunication Systems*, 49(2):207–217, 2012.

26. V. Berk, A. Giani, and G. Cybenko. Detection of covert channel encoding in network packet delays. Technical Report TR2005-536, Department of Computer Science, Dartmouth College, November 2005. http://www.ists.dartmouth.edu/library/149.pdf.

27. S. Gianvecchio, H. Wang, D. Wijesekera, and S. Jajodia. Model-based covert timing channels: automated modeling and evasion. In *Recent Advances in Intrusion Detection (RAID) Symposium*, September 2008.

28. S. H. Sellke, C.-C. Wang, S. Bagchi, and N. B. Shroff. Covert TCP/IP timing channels: theory to implementation. In *Proceedings of the 28th Conference on Computer Communications (INFOCOM)*, April 2009.

29. S. Gianvecchio and H. Wang. Detecting covert timing channels: an entropy-based approach. In *Proceedings of the 14th ACM Conference on Computer and Communication Security (CCS)*, November 2007.

30. S. Zander, G. Armitage, and P. Branch. Stealthier interpacket timing covert channels. In *IFIP Networking*, May 9–13, 2011.

31. Y. Liu, D. Ghosal, F. Armknecht, A.-R. Sadeghi, S. Schulz, and S. Katzenbeisser. Hide and seek in time: robust covert timing channels. In *Proceedings of the 14th European Symposium on Research in Computer Security*, September 2009.

32. G. Shah, A. Molina, and M. Blaze. Keyboards and covert channels. In *Proceedings of the USENIX Security Symposium*, August 2006.

33. A. Houmansadr, N. Kiyavash, and N. Borisov. RAINBOW: a robust and invisible non-blind watermark for network flows. In *Proceedings of the 16th Annual Network & Distributed System Security Symposium (NDSS)*, February 2009.

34. M. H. Kang and I. S. Moskowitz. A pump for rapid, reliable, secure communication. In *Proceedings of the ACM Conference on Computer and Communications Security (CCS)*, pp. 119–129, 1993.

35. N. Ogurtsov, H. Orman, R. Schroeppel, S. O'Malley, and O. Spatscheck. Experimental results of covert channel limitation in one-way communication systems. In *Proceedings of the Symposium on Network and Distributed System Security (SNDSS)*, February 1997.

36. H.-G. Eßer and F. C. Freiling. Kapazitätsmessung eines verdeckten Zeitkanals über HTTP. Technical Report TR-2005-10, Universitt Mannheim, 2005. http://bibserv7.bib.uni-mannheim.de/madoc/volltexte/2005/1136/pdf/tr_2005_10.pdf (in german).

37. S. Li and A. Ephremides. A network layer covert channel in ad-hoc wireless networks. In *Proceedings of the First IEEE Conference on Sensor and Ad Hoc Communications and Networks (SECON)*, pp. 88–96, October 2004.

38. S. D. Servetto and M. Vetterli. Communication using phantoms: covert channels in the internet. In *Proceedings of the IEEE International Symposium on Information Theory (ISIT)*, June 2001.

39. A.-R. Sadeghi, S. Schulz, and V. Varadharajan. The silence of the LANs: efficient leakage resilience for IPsec VPNs. In *Computer Security: ESORICS 2012*, Vol. 7459 of *Lecture Notes in Computer Science*, pp. 253–270, 2012.

40. C. Krätzer, J. Dittmann, A. Lang, and T. Kühne. WLAN steganography: a first practical review. In *Proceedings of the 8th ACM Multimedia and Security Workshop*, September 2006.

41. D. Kundur and K. Ahsan. Practical Internet steganography: data hiding in IP. In *Proceedings of the Texas Workshop on Security of Information Systems*, April 2003.

42. K. Ahsan and D. Kundur. Practical data hiding in TCP/IP. In *Proceedings of the ACM Workshop on Multimedia Security*, December 2002.

43. R. C. Chakinala, A. Kumarasubramanian, R. Manokaran, G. Noubir, C. P. Rangan, and R. Sundaram. Steganographic communication in ordered channels. In *Proceedings of the 8th International Workshop on Information Hiding*, pp. 42–57, July 2006.

44. X. Luo, E. W. W. Chan, and R. K. C. Chang. Cloak: a ten-fold Way for reliable covert communications. In *Proceedings of European Symposium on Research in Computer Security (ESORICS)*, September 2007.

45. A. El-Atawy and E. Al-Shaer. Building covert channels over the packet reordering phenomenon. In *Proceedings of the 28th Annual IEEE Conference on Computer Communications (INFOCOM)*, 2009.

46. T. M. Dogu and A. Ephremides. Covert information transmission through the use of standard collision resolution algorithms. In *Proceedings of the Third International Workshop on Information Hiding (IH)*, pp. 419–433, September 1999.

47. S. Li and A. Ephremides. A covert channel in MAC protocols based on splitting algorithms. In *Proceedings of the Wireless Communications and Networking Conference (WCNC)*, pp. 1168–1173, March 2005.

48. S. J. Murdoch. Hot or not: revealing hidden services by their clock skew. In *Proceedings of the 13th ACM Conference on Computer and Communications Security (CCS)*, pp. 27–36, November 2006.

49. S. Zander, P. Branch, and G. Armitage. Capacity of temperature-based covert channels. *IEEE Communications Letters*, 15(1):82–84, 2010.

50. S. Zander and S. J. Murdoch. An improved clock-skew measurement technique for revealing hidden services. In *Proceedings of Usenix Security*, July/August 2008.

51. R. Dingledine, N. Mathewson, and P. F. Syverson. Tor: The second-generation onion router. In *Proceedings of the 13th USENIX Security Symposium*, August 2004.

52. A. Hintz. Covert channels in TCP and IP Headers, 2003. http://www.defcon.org/images/defcon-10/dc-10-presentations/dc10-hintz-covert.ppt.

53. W. Mazurczyk and K. Szczypiorski. Steganography of VoIP streams. In I. R. Meersman and Z. T., editors, *Proceedings of the 3rd International Symposium on Information Security (IS'08), Monterrey, Mexico*, Vol. 5332 of *Lecture Notes in Computer Science*, pp. 1001–1018. Springer, Berlin, 2008.

54. W. Mazurczyk, M. Smolarczyk, and K. Szczypiorski. Retransmission steganography and its detection. *Soft Computing*, 15(3):505–515, 2011.

55. W. Mazurczyk and J. Lubacz. LACK: a VoIP steganographic method. *Telecommunication Systems: Modelling, Analysis, Design and Management*, 45(2-3):153–163, 2010.

56. W. Mazurczyk, J. Lubacz, and K. Szczypiorski. On steganography in lost audio packets. *International Journal of Security and Communication Networks*, 2012, doi: *arXiv:1102.0023*.

57. W. Mazurczyk. Lost audio packets steganography: a first practical evaluation. *International Journal of Security and Communication Networks*, 5(12):1394–1403, 2012.

58. M. Hamdaqa and L. Tahvildari. ReLACK: a reliable VoIP steganography approach. In *2011 Fifth International Conference on Secure Software Integration and Reliability Improvement (SSIRI)*, pp. 189–197, June 2011.

59. C. Arackaparambil, Y. Guanhua, S. Bratus, and A. Caglayan. On tuning the knobs of distribution-based methods for detecting VoIP covert channels. In *45th Hawaii International Conference on System Science (HICSS '12)*, pp. 2431–2440, January 2012.

60. S. Na and S. Yoo. Allowable propagation delay for VoIP calls of acceptable quality. In *Proceedings of the AISA 2002*, Vol. 2402/2002, pp. 469–480. Springer, Berlin, 2002.

61. H. Schulzrinne, S. Casner, R. Frederick, and V. Jacobson. RTP: a transport protocol for real-time applications, July 2003.

62. T. Friedman, R. Caceres, and A. Clark. RTP control protocol extended reports (RTCP XR), November 2003.

63. W. Mazurczyk, M. Smolarczyk, and K. Szczypiorski. On information hiding in retransmissions. *Telecommunication Systems*, 52(2):1113–1121, 2013.

64. J. Zhai, G. Liu, and Y. Dai. An improved retransmission steganography and its detection algorithm. In *2011 Third International Conference on Multimedia Information Networking and Security (MINES)*, pp. 628–632, November 2011.

65. S. Rewaskar, J. Kaur, and F. D. Smith. A performance study of loss detection/recovery in real-world TCP implementations. In *Proceedings of the ICNP 2007*, pp. 256–265, 2007.

66. C. Chen, M. Mangrulkar, N. Ramos, and M. Sarkar. Trends in TCP/IP retransmissions and resets. Technical Report, University of California, San Diego, 2001.

CONTROL PROTOCOLS FOR RELIABLE NETWORK STEGANOGRAPHY

> It is the unexpected sequences of events that lead to protocol failures, and the hardest problem in protocol design is precisely that we must try to expect the unexpected.
>
> —*Gerald J. Holzmann*

Early network steganography research was focused on the goal to transfer a steganogram from a sender to a receiver. With the increasing number of use cases explained in Chapters 1 and 2, novel requirements for network steganography arose. These requirements especially include but are not limited to the need for a reliable steganographic communication. This chapter deals with so-called *control protocols* that serve as enablers for feature-rich network steganography. In referenced work, control protocols are also referred to as *micro-protocols*. For better readability, we decided not to use the term *micro-protocol* and solely rely on the term *control protocol* instead.

4.1 STEGANOGRAPHIC CONTROL PROTOCOLS

Before we can define a steganographic control protocol, we need to understand the idea of a communication protocol.

Information Hiding in Communication Networks: Fundamentals, Mechanisms, Applications, and Countermeasures,
First Edition. Wojciech Mazurczyk, Steffen Wendzel, Sebastian Zander, Amir Houmansadr, and Krzysztof Szczypiorski.
© 2016 by The Institute of Electrical and Electronics Engineers, Inc. Published 2016 by John Wiley & Sons, Inc.

Communication Protocol   Communication protocols *regulate the communication between distributed processes in a computer network* [1].

Steganographic Control Protocol   A steganographic control protocol is a communication protocol that is embedded into a steganographic carrier; it regulates the communication between distributed steganographic processes.

Control protocols exist for both storage and timing channels. Control protocols placed within *storage* channels are embedded into storage locations of carriers. Storage channel-based control protocols are known to comprise a large variety of features. For *timing* channels, control protocols are represented through redundant information within temporal behavior of network data and typically provide fewer features as in the case of storage channels. Both types of control protocols will be discussed in this chapter, but as most work is available on storage channels, the majority of sections will focus on these.

The first terminology for control protocols was introduced in 2012 [2], but a more detailed terminology that is conforms to Chapter 2 will be presented here.

As shown in Figure 4.1, both the transferred payload and the control protocol's data are embedded into the utilized subcarriers of a packet (represented as hatched area in the network packet), forming a logically combined area called *cover area* [3].

The control protocol can contain multiple components (represented as puzzle pieces in Figure 4.1). In general, a control protocol consists of mandatory meta information that describes the header itself (e.g., flags that determine the presence of optional header components) and the header parameters (e.g., an optional header component).

4.1.1   Features of Control Protocols

Control protocols enable manifold features for network steganography [4]. First of all, control protocols were introduced to provide reliability for the transfer of steganograms by introducing acknowledgment flags and sequence numbers. Control protocols can also be used to create, manage, and end sessions within covert data transfers.

Another important feature implementable with control protocols are overlay networks. Overlay networks are networks on top of the existing networks, such as P2P file-sharing networks on top of the IP network, that can have their own addressing and routing concepts. Similarly, steganographic control protocols can enable an own addressing scheme and even functionality of dynamic routing protocols and mobile overlay networks [3].

It can be necessary to discover both, steganographic peers in the overlay network and the currently applicable steganographic methods that can be used to communicate with a particular peer. Both discovery functions, peer discovery and steganographic

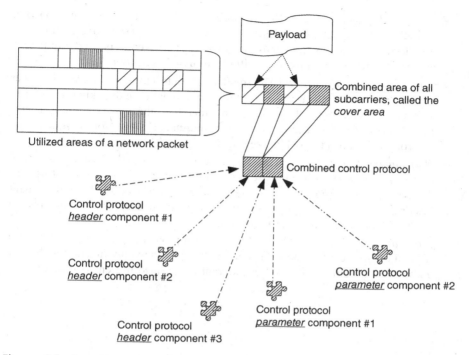

Figure 4.1. Control protocol terminology showing the embedding of all control protocol components into subcarriers, which are combined to form the cover area.

methods discovery, can be realized on the basis of control protocols. As steganographic methods may be switched or placed in different subcarriers, control protocols can

- initiate the switch of subcarriers to be used;
- initiate a switch of the steganographic method itself.

A steganographic control protocol provides *adaptiveness*, when it allows the channel to adjust itself to fit new requirements (e.g., switching a hiding method, discovering a new peer, or using/providing dynamic routing).

Since control protocols can contain their own protocol version numbers and version numbers of the used steganographic software, they enable backward-compatible overlays as well as upgrades of the steganographic infrastructure. This also means steganograms can be used to transfer new versions of the steganographic software.

4.1.2 Requirements for Control Protocols

As pointed out in [4], the design of control protocols is linked to specific requirements:

- *Low-profile control protocol embedding:* In order to minimize the anomalies caused by the signaling (placement) of the control protocol, one requirement is a design that adapts the control protocol to the subcarriers used. Optimized engineering for control protocols will be discussed in Section 4.3.

- *Control protocol header size:* The smaller the control protocol header, the fewer the bits hidden in a subcarrier, causing fewer anomalies and making it harder to detect the control protocol operation (and, therefore, the covert channel). Therefore, a small size of the header is another requirement. Means for minimizing the control protocol header will be presented in Section 4.3.

- *High number of features per header bit:* The more efficient the design of a control protocol, the more the features realized per bit of the control protocol header. Achieving a high *features per header bit* value is therefore another requirement besides minimizing the header size. In other words, header size and features are conflicting goals, as shown in Figure 4.2.

- *Minimal dependence on TCP/IP:* Most control protocols were designed for and depend on the TCP/IP protocol suite, for example, they only function with IPv4 addresses. As future network steganography may also use embedding of non-

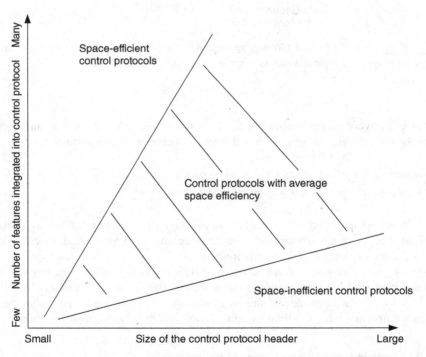

Figure 4.2. Optimization problem for control protocols: header size and feature spectrum are conflicting requirements.

Magic	Dst. IP	Dst. Port	State	Ack#	Length	Seq#	ID
32 bit	32 bit	32 bit	32 bit	32 bit	32 bit	16 bit	16 bit

Figure 4.3. Ping Tunnel's control protocol header. (Reproduced from [4] with permission of Springer.)

TCP/IP environments (e.g., CPS environments, cf. [5]), it is another requirement to design control protocols in a way that is independent of the utilized (sub)carrier.

- *Upgradability:* As with other network protocols, control protocols for network steganography can be subject to changes. This means that newer protocol versions may be introduced. For this purpose, version numbering and the so-called *network environment learning* (NEL) *phase* (cf. Sections 4.4 and 4.3.2) can be supported by a control protocol. This allows for upgrading network steganographic overlay networks, e.g. to introduce new control protocol features. Another requirement is thus the upgradability of control protocols.

- *Handling of noncovert input:* As network steganographic communications take place within environments that allow legitimate communications at the same time, another requirement of a control protocol is to identify and ignore noncovert data.

4.1.3 Existing Control Protocols

We will now discuss the existing control protocols for network information hiding, based on [4].

4.1.3.1 Ping Tunnel.

The tool *Ping Tunnel* was developed by Daniel Stødle in 2004 [6] and its development still continues as the tool is well known in the hacking community and its main purpose to bypass firewalls is making it an attractive tool, for example, for penetration testing. Ping Tunnel's basic functionality is to transfer hidden data within the ICMP Echo request and Echo reply payload in a reliable manner, that is, Ping Tunnel is capable of handling lost packets by resending nonacknowledged packets.

The Echo payload serves as subcarrier in which the covert payload as well as the control protocol is hidden. The first part of the Echo payload is used for the control protocol header, followed by the actual hidden payload.

The Ping Tunnel header is shown in Figure 4.3 and starts with a 4 byte magic number used to differentiate between ICMP Echo packets that belong to Ping Tunnel traffic and those who do not belong to Ping Tunnel traffic. Therefore, the magic number is always set to the value $0xD5200880$. This constant magic number makes Ping Tunnel easy to detect and block.

The magic number is followed by a 4 byte destination IP address and a 4 byte destination port of which only 16 bits are used, as the size of a TCP/UDP port number is 16 bits. Both values are required to forward traffic to the destination using a Ping Tunnel proxy. These two addressing fields follow a 4 byte state information field that is used to indicate the message type (initiation of new proxy session, data forwarding,

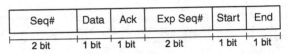

Figure 4.4. The header of the protocol presented by Ray and Mishra. (Reproduced from [4] with permission of Springer.)

acknowledgment, close of connection, and authentication message) and sender specification that specifies whether the message was sent by a client or by a proxy.

It follows a 4 byte acknowledgment number of which only 2 bytes are used, a 4 byte payload length field, a 2 byte sequence number, and a 2 byte identifier to handle simultaneous connections.

4.1.3.2 Control Protocol by deGraaf et al.

A rather simple control protocol was proposed by deGraaf et al. in 2005 [7]. The control protocol aims at preventing reordering of hidden data transmissions while using UDP. A reordering is possible if one packet overruns another one due to a different routing path. The control protocol is hidden into the UDP destination port field [7] by splitting the port into a data part and a sequence number. The protocol is mentioned here as it is—although introduced 1 year after Ping Tunnel—the first control protocol discussed within the research community. deGraaf et al. designed the control protocol for a port knocking-based covert channel.

4.1.3.3 Control Protocol by Ray and Mishra.

Ray and Mishra released a control protocol in 2008 [8]. Like Ping Tunnel, this protocol is placed into the ICMP Echo payload.

The header of the control protocol is shown in Figure 4.4 and features a sequence number and an expected sequence number that functions as an acknowledgment number (2 bits each). One flag indicates that payload is attached and another flag indicates that the packet contains an acknowledgment.

The field *expected sequence number* contains the sequence number of the last received packet, e.g. if the field contains the number *01*, the packet with the sequence number *1* was successfully received, but the sequence number *10* (2) is expected for the next packet.

The two bit sequence number is sufficient for a covert channel communication [8]. On first sight and in comparison to Ping Tunnel's 16 bit sequence number, the 2 bit sequence number of the Ray–Mishra protocol is small. But since the Ray–Mishra protocol transmits new packets only after the latest packet was received and acknowledged, sequence numbers cannot be used for two packets at the same time. In other words, a *stop-and-wait automatic repeat request* (ARQ) protocol is implemented. However, waiting for the acknowledgment of a packet before sending out the next packet is a slow process. Since ARQ only requires a 1 bit sequence number but the protocol provides a 2 bit sequence number, an improved variant of ARQ could be used where multiple packets can be sent sequentially before an acknowledgment is received. Therefore, Ray and Mishra propose the improved algorithms *Go-back-n ARQ* and *selective repeat ARQ*

[8] but these can lead to more retransmissions of packets in case of ICMP rate limiting,[1] which can raise more attention.

The last two bits are used to specify whether a covert message starts or ends with the current packet.

4.1.3.4 Covert File Transfer Protocol (CFTP).
In 2010, Trabelsi and Jawhar published work on a *covert file transfer protocol* (CFTP) [10]. CFTP is embedded within the IP record route option; it allows to transfer covert messages and to upload, download, and list hidden files, similarly like the FTP protocol. To this end, CFTP provides a file hosting service—the CFTP server—that is accessible by the CFTP clients. CFTP additionally implements session management and reliability, but in comparison to the other control protocols, it also provides authorization. For instance, a client can be allowed to upload files on the CFTP server, but the server could forbid the same client to list the files located on the server. These permissions are indicated by four flags (privilege to list files, download files, upload files, and to send short messages). Furthermore, additional flags indicate whether a packet is a short message, a file, or a part of a file, and whether a packet is a retransmitted packet or not.

4.1.3.5 Hybrid Approach Using Digital Watermarking.
Mazurczyk and Kotulski developed a control protocol based on a hybrid hiding approach in 2006 [11]. The control protocol is not only relevant for a hidden communication but also serves the purpose of extending the capabilities of RTP without requiring additional header fields.

The control protocol combines a covert channel with digital watermarking. A 6 bit control protocol header is embedded into unused bits of the IP, UDP, and RTP header. The header describes control protocol parameters inside a watermark embedded into the payload of a VoIP message. The advantage of the hybrid approach is to keep the space required for the major components of the control protocol small by placing parts of the control protocol data into the watermark.

The first four bits of the control protocol describe the information contained in the watermark, which are analogous to the RTCP protocol (authentication or integrity parameters, interarrival jitter, NTP timestamp or RTP timestamp, just to mention a few). The fifth bit signals the side of the bidirectional communication (sender or receiver). The sixth bit indicates whether the packet represents a beginning parameter indicated by the first four bytes or whether the watermark contains continuing data of a parameter.

4.1.3.6 TrustMAS and the Smart Covert Channel Tool (SCCT).
TrustMAS (*Trusted Multiagent Systems*) is a dynamic routing protocol to build covert overlay networks and was developed by Szczypiorski et al. [12]. TrustMAS provides peer discovery and routing update functionality. To realize dynamic routing, the random walk algorithm is used. To decide whether a message is directly forwarded to a destination or sent to another hop, a coin-flip decision is performed—a task repeated until the message

---

[1] With ICMP rate limiting, e.g., provided by modern CISCO devices and Linux, the number of ICMP messages of the same type per time slot can be limited [9].

reaches its destination. The algorithm scales well for a high number of participants and is the first applied to steganographic overlay routing.

In 2012, Backs et al. presented a size-optimized control protocol that allows the configuration of dynamic routing in covert channel overlay networks. The protocol improves over TrustMAS by selecting routes linked to a high covertness [13]. Their so-called *smart covert channel tool* (SCCT) implements a routing algorithm based on OLSR (optimized link state routing) to achieve a high stealthiness for the propagation of routing information. Four routing messages (requesting peer tables, sending peer tables, transferring topology graphs, and performing routing updates) are provided.

SCCT provides reliability for overlay networks by handling the integration of new participants, the disconnection and crash management of participants, and the prohibition of explicit communication between two selected participants, e.g. to prevent direct communication over a link that can only provide very low covertness. In order to maximize the covertness of a link between two participants, SCCT uses a *quality of covertness* (QoC) measure for the stealthiness of a routing path. SCCT can also utilize so-called *drones*, that is, systems used as hops for the overlay routing that are not aware of any covert communication. For this reason, drones are only able to forward data, but unable to interpret covert messages. Typical drones are public web services allowing remote website requests, such as Google Translator.

The header of the control protocol is highly dynamic and only includes the header components required for a particular routing message. The authors therefore introduce the concept of *status updates*, which enabled the dynamic header design and which will be discussed in Section 4.3. Moreover, the control protocol of SCCT is designed to be integrated into various carriers.

4.1.4 Comparison of Existing Control Protocols

Wendzel and Keller compared existing control protocols [4], and their comparison serves as a basis for this section.

Starting the comparison with a view on the features linked to different OSI layers, Wendzel and Keller noticed that only one of the protocols, namely, CFTP, provides functionality of the application layer [4]. Features of the presentation layer, such as determining the cryptographic authenticity of the covert data, are only implemented by the hybrid approach of Mazurczyk and Kotulski [11]. Session layer functionality (OSI layer 5) is integrated by the majority of control protocols and so are transport layer features, especially reliability. Network layer functionality is supported only by Ping Tunnel and SCCT. None of the inspected control protocols comprises functionality of the data link or physical layer because, being embedded into higher level protocols, they do not require to support these.

In contrast to the OSI layer-based view, a second comparison of the known control protocols was conducted with regard to the requirements of Section 4.1.2 [4] (e.g., hiding placement or optimized header size). The known hiding methods applied for control protocols are simplistic in the sense that no specific engineering methods are applied for optimized embedding. However, the protocol by Ray and Mishra [8] and the hybrid approach by Mazurczyk and Kotulski [11] possess size-optimized headers.

SCCT is the only protocol using a control protocol engineering method to minimize header size, namely, status updates (Section 4.3.2).

Comparing the *features per header bit* value of all protocols is not directly feasible as SCCT and the hybrid approach by Mazurczyk and Kotulski have a dynamic protocol structure. The control protocol by Ray and Mishra achieves the highest value $(0.75)^2$ while Ping Tunnel and the protocol by deGraaf et al. [7] result in the lowest values (both ≤ 0.1).

Only Ping Tunnel and SCCT depend on the TCP/IP protocol suite (due to their addressing scheme) and none of the protocols besides SCCT can be directly upgraded (even for SCCT, this feature is only proposed). Out of band data can be handled only by simple reliability features by most of the protocols—a problem that later resulted in attacks on control protocols (Section 8.1.2).

4.2 DEEP HIDING TECHNIQUES

Network steganographic transmissions can switch their utilized carriers, subcarriers, and hiding methods to increase covertness and to improve the capability to bypass protection measures [3,14,15]. These techniques are also referred to as *deep hiding techniques* (DHTs) [16] and can aggravate a forensic steganalysis even if all traffic is recorded since all n (sub)carriers must be analyzed; moreover, if only m of the n (with $m \leq n$) (sub)carriers are found by a warden, the covert channels in the remaining $n - m$ (sub)carriers still remain undetected [3].

Switching (sub)carriers usually means switching the used hiding methods. If reliability, error detection, or error correction must be provided across different carriers or subcarriers, a control protocol or a suitable coding must be used. Especially, a control protocol can embed sequence numbers into all transmissions to reorder jumbled packet sequences at the receiver side [15]. As shown in [17], DHTs can also affect the steganographic cost.

Different approaches for switching the used hiding method and (sub)carrier exist. While a first method for carrier switching occurred in the late 1990s in the hacking community [18], various advanced approaches were presented by two research groups between 2007–2012, namely, Wendzel and Keller [3,15,19] as well as Mazurczyk, Szczypiorski, Fraczek, and Jankowski [14,16,20]. The majority of these DHTs were published simultaneously (with a time-delta of only 2 weeks by both research groups). In 2012, countermeasures for selected DHTs were developed by Wendzel, Keller, and Zander [21,22] (cf. Sections 8.4.5 and 8.6.7). This section merges the known research literature's terms and concepts to provide the reader a unified understanding of these. In particular, the following techniques are known:

1. Switching of the carrier, that is, a *protocol switching covert channel* (PSCC). Three types of PSCC exist:

---

[2] The protocol's *features per header bit* value was later optimized by work of Backs et al. [13].

Type 1: The earliest realization of a covert channel known to use two network protocols for the embedding of secret data was developed within the hacking community in 1997. The implementation is known as *LOKI2* [18]. LOKI2 could only hide traffic in one protocol a time, and its user needed to initiate the protocol switch manually. LOKI2 can be referred to as a type of *protocol hopping covert channel* (PHCC) [15] or *steganogram scattering* (SGS) method [16]. These channels transfer hidden data via at least two different network protocols between sender and receiver. In general, they can also utilize different carriers simultaneously and in a transparent manner, that is, not triggered by the end user [3,15,16]. Moreover, the sender as well as the receiver can be a distributed system, that is, multiple hosts can send and receive the PSCC traffic [3,16]. We discuss the optimization of PSCCs in Sections 4.4.1 and 4.4.2.

Type 2: The concept of a *protocol channel* (PC) was published in 2008 [19]. In comparison to techniques of type 1, a PC does not "embed" hidden information into a packet, instead the *protocol* used for a packet represents the hidden information itself. If n protocols are used by a PC, $\log_2(n)$ bits can be transferred per packet. Both PSCC types, 1 and 2, are visualized in Figure 4.5.

Type 3: Instead of sending hidden payload over multiple carriers, a covert channel can explicitly utilize the relationships between different network protocols. This type of covert channel is called *inter-protocol steganography* (IPS) [14,16]. For instance, *PadSteg* may only hide data in the padding of an Ethernet frame if a given protocol of an upper communication layer is embedded [14].

2. Switching of the subcarrier, that is, *steganogram hopping* (SGH) [3,16]. In comparison to a PSCC, a SGH switches between subcarriers within the same carrier. For instance, an IPv4-based SGH could switch between the utilization of the least significant TTL bit and the reserved bit. As in case of different carriers, the utilization of different subcarriers can be optimized [3].

3. Fraczek et al. proposed the concept of *multilevel steganography* (MLS) [16,20]. At least two hiding methods are required to perform MLS. The first method hides data within a carrier, while the second hiding method uses the first as a carrier to hide its data. Additional layers can be added on demand.

4. Moreover, hybrid forms of the above-mentioned techniques are feasible [16].

4.3 CONTROL PROTOCOL ENGINEERING

When designing a control protocol for a covert storage channel, methods available from the area of *protocol engineering* can be applied to design, verify, and test the protocol. Research on such methods was performed for decades and the

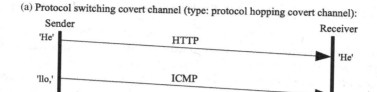

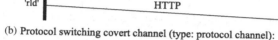

Figure 4.5. Two types of PSCCs. (a) Protocol hopping covert channel using two protocols (hidden data are embedded into storage attributes). (b) Protocol channel using four protocols (hidden data are represented by the protocol itself). (Reproduced from [21] with permission of Iaria.)

available textbooks, especially [23], provide comprehensive information on the subject, including a coverage of formal methods such as Petri nets and formal grammar.

For control protocols, it is essential to discuss protocol engineering means that were developed for the specific needs of steganographic communications, that is, a control protocol should be designed and placed into a carrier in a way that it causes as few anomalies as possible. First, the number of bits needed by the control protocol should be as small as possible—the fewer the data dedicated for a control protocol header, the fewer the carrier bits manipulated. Second, the cover carrier should follow the rules of the utilized carrier, that is, it should not cause behavior of the carrier that is not foreseen by the communication standard (e.g., by an Internet protocol standard defined by an RFC). We discuss these requirements separately after introducing elementary aspects for the placement of a control protocol header.

4.3.1 Elementary Aspects for Embedding a Control Protocol

A fundamental aspect of a control protocol is the general limitation of its header size. An adaptive control protocol can be embedded into different carriers. These different carriers provide varying space for the control protocol. Given a number of carriers $C_1, \cdots, C_n$, each carrier C_i provides a subcarrier space of size s_i for the placement of a steganogram [3]. If a control protocol is to be designed that should fit into all possible carriers, then the size of the control protocol (s_{header}) can have a maximum size of

$$s_{\text{header}} = \min(s_1, \ldots, s_n).$$

However, transferring solely control protocol headers can be considered useless in most scenarios and, thus, at least a small space for payload (s_{payload}) should be reserved, resulting in

$$s_{\text{header}} = \min(s_1, \ldots, s_n) - s_{\text{payload}}$$

and a minimum space requirement ($s_{\text{min}} = s_{\text{header}} + s_{\text{payload}}$) that is valid for all carriers [3]. However, as some carriers provide more space than others, the particular space for payload can vary:

$$s_{i_{\text{payload}}} = s_i - s_{\text{header}}.$$

4.3.2 Status Updates: Dynamic Control Protocol Headers

An approach to minimize the size of control protocol headers was presented by Backs et al. in 2012 [13] and is called *status updates*. The concept of status updates is based on two ideas.

First, status updates utilize a simple compression technique already used by the *compressed serial line interface protocol* (CSLIP) [24]. CSLIP transfers a bitmask to indicate present header fields, followed by the indicated header fields themselves. As long as the bitmask is smaller than the combined size of nonpresent header fields, the transferred packets become smaller than in the case of full headers. A similar concept was already used earlier by the hybrid watermarking approach (cf. Section 4.1.3.5) in which a header field indicated the presence of parameters in the watermark.

Second, status updates apply the concept of succeeding header fields that can be found in the IPv6 protocol. The *next header* field in IPv6 indicates the presence and type of a following header. This idea makes the CSLIP concept dynamic as headers can occur in almost arbitrary order and even multiple times.

4.3.2.1 How Status Updates Work.
In case of status updates, there is just one field indicating the following header field (similar to IPv6) and the field uses a CSLIP-like bitmask to inform the receiver about the type of the header. This field is called *type of update* (ToU) field. The ToU field is followed by the actual header field, which again is followed by a ToU field that indicates the following header type (or the end of the header list). Status updates can thus be understood as if an IPv6-like protocol's

ToU lookup table:

00 - SET SOURCE ADDRESS
01 - SET DESTINATION ADDRESS
10 - PAYLOAD FOLLOWS
11 - END OF UPDATES

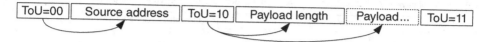

Figure 4.6. The concept of status updates.

header would solely consist of the *next header* field. The concept is visualized for an exemplary two bit ToU in Figure 4.6.

However, it is not necessary to define a ToU for each particular header component, that is, a single ToU can be used for multiple header components, which decreases the required space for ToU values in the control protocol header, as shown for the ToU 10 in Figure 4.6.

The technique is called *status updates* because sender and receiver keep track of different states of a covert connection between two peers. If, for instance, one peer is asked to forward data to another peer, the destination address of the payload is indicated via a ToU that sets the new destination address. All following payload does not need to specify the destination again (as the status is kept), which removes the necessity to transfer the destination address again until it is changed.

4.3.2.2 Advantages of Status Updates.

Status updates decrease the size of the control protocol header (i.e., optimize the *features per header bit* metric). Header fields indicated by status updates can also occur multiple times per packet to reduce the number of required packets per transaction (Figure 4.7) and can also occur in a freely definable order (Figure 4.6); they can moreover combine payload and commands of succeeding packets within a single packet if the available space s_i is large enough and can decrease the packet count for a transfer (also visualized in Figure 4.7). To this end, the areas used for the control protocol header and for the covert payload can be fragmented (cf. Figure 4.1).

Another advantage is that existing status updates can be extended to allow the *evolution of a control protocol*. For this at least one possible status update value must be still definable within the existing ToU list (i.e., if n bits are available to define ToUs, then 1 ToU of the possible 2^n ToU values must still be available to allow the integration of an additional ToU). Optionally, the additional ToU can be a *meta-ToU*, that is, a ToU followed by extended ToU values. For instance, if only one possible ToU value is left for definition, this ToU value can be defined as a meta-ToU value. A meta-ToU value is followed by a n bit ToU value extension that represents new ToU values and optionally leaves space for an additional extension. ToUs could be followed by a version number to handle multiple versions of control protocols.

Using the status update approach, the previously introduced control protocol by Ray and Mishra (cf. Section 4.1.3.3) can be designed in a more space-efficient way [13].

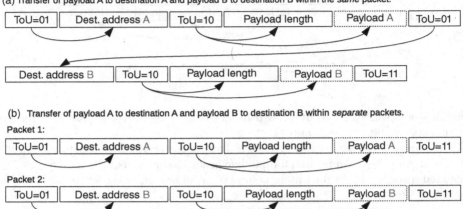

ToU lookup table: 00 - SET SOURCE ADDRESS
01 - SET DESTINATION ADDRESS
10 - PAYLOAD FOLLOWS
11 - END OF UPDATES

(a) Transfer of payload A to destination A and payload B to destination B within the *same* packet.

(b) Transfer of payload A to destination A and payload B to destination B within *separate* packets.

Packet 1:

Packet 2:

Figure 4.7. (a) A ToU occurs multiple times within one packet to reduce the overall number of packets and header bits required for a transaction. (b) The same data are transmitted using two packets, that is, the feature of allowing multiple occurrences for a ToU per packet is not used.

Backs et al. additionally implemented a dynamic routing protocol for covert channel networks on the basis of status updates [13].

4.3.3 Design of Conforming Control Protocols

Besides reducing the size of the control protocol, there is another approach to improve the stealthiness of a control protocol. In [2], Wendzel and Keller present an engineering approach that first designs a control protocol in a way that it does not violate the rules of the utilized carrier. In other words, nonforeseen protocol states in the utilized carrier are not caused by the operation of the control protocol when it is placed inside the carrier. Second, the control protocol placement is optimized so that the placement causes as few anomalies as possible (e.g., unusual occurrence rates of header field values in the cover carrier). Step 2 is necessary because bits of the control protocol can have different occurrence rates compared to bits of the utilized carrier.

We modified the approach presented originally in [2] to fit into the terminology presented in Figure 4.8.[3] Solid arrows indicate the flow of the engineering process

---

[3] Originally, the terms *micro-protocol* (instead of control protocol), *cover protocol* (instead of cover carrier), and *underlying protocol* (instead of (sub)carrier) were used.

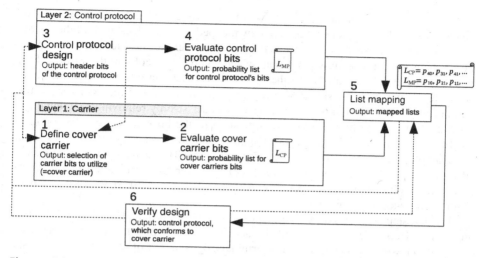

Figure 4.8. Control protocol engineering approach. (Reproduced from [2] with permission of Springer.)

and dashed arrows represent re-engineering paths that are used to optimize the control protocol design in an iterative process.

The first 4 steps of the approach are split into two layers: one layer in which the design process focuses on the (sub)carrier and one layer in which the design process focuses on the control protocol.

The process starts with the carrier layer, where suitable subcarriers are selected for the placement of the control protocol. This logically combined area is called the *cover carrier* and represented by the previously introduced s_i for the carrier C_i. It is important to emphasize again that the control protocol can be split over an arbitrary number of subcarriers (cf. Figure 4.1).

Step 2 evaluates the probability of all bit values of the cover carrier. For instance, a bit i in the subcarrier is set with $p(i = 1) = 0.85$ and unset with $p(i = 0) = 0.15$. This step can—for instance—be based on the evaluation of traffic recordings but also on estimations. In simple cases, protocol designers could also use rough categorizations such as *low*, *medium*, and *high* for bit occurrence rates. The output of this step is a list of cover carrier bits with their particular occurrence rates.

Step 3 concentrates on the control protocol itself. In this step, the control protocol header is designed, that is, all necessary bits and functions are defined and the protocol specification is written down. This step must be accomplished under the given condition of step 1 as the definition of the cover carrier defines the number of available bits in the cover carrier and the control protocol must fit into these cover carrier bits.

Step 4 is similar to step 2, but evaluates the occurrence rates of control protocol bits based on simulation or estimation. The output of this step is a list of control protocol bits with their particular occurrence rates.

Both layers are combined in step 5. Bits of the cover carrier and the control protocol, which have similar occurrence rates, are now mapped. In the end, each bit of the control protocol must be mapped to exactly one bit of the cover carrier. Steps 2, 4, and 5 are thus responsible for a stealthy placement of the control protocol into the carrier. If the mapping in step 5 is unsatisfying, the protocol designer can either change the cover carrier (e.g., choose other subcarriers or increase the size of a subcarrier) or modify the control protocol design (e.g., decrease the header size by removing functionality). A detailed discussion regarding the optimization of the bit mapping can be found in [2]. In [4], the authors stated that bit mappings should depend on the carrier's state. For example, when TCP serves as a carrier and a new TCP connection is established, different TCP flags are more likely to be set as within the termination phase of the carrier. If the bit mapping is performed equally for all states of the TCP protocol, it may result in an nonoptimal mapping.

Finally, step 6 verifies the conformance of the control protocol. To this end, control protocol and cover carrier *both* are defined using a separate formal grammar $G = (V, \sum, P, S)$, where V is the set of nonterminals, $\sum$ is the set of terminals, P is the set of productions, and $S \in V$ is the start symbol [25].

In this approach, each of the two grammar's terminal symbols represent the bits of the particular protocol header, with both grammars having the same terminal symbols to represent the same bits (achieved by the mapping in step 5). The production rules represent the rules of the protocol, for example, allowing to set some bits at the same time, while others are not allowed to occur at the same time.

Finally, the protocol designer verifies whether the language $L(G_{CP})$ of the control protocol's grammar is part of (or equal to) the language produced by the cover carrier's grammar $L(G_{CC})$. This verification is accomplished using a language inclusion test $L(G_{CP}) \subseteq L(G_{CC})$, which can be automated if a regular or context-free grammar is used to generate $L(G_{CP})$ and a regular grammar is used to generate $L(G_{CC})$.

Example: We select a two bit field in an imaginary network protocol as our cover carrier. The protocol allows a sender to transfer payload to a receiver and allows the receiver to send acknowledgments for received payload. The sender can optionally request an acknowledgment for attached payload and can also indicate an error.

The two bits a and b of the cover carrier represent the following functionality:

- Payload is transferred if bit a is set.
- An acknowledgment for the attached payload is requested if bit b is set.
- An error is indicated if neither a nor b is set.

It makes little sense to transfer a packet in which bit a is unset but bit b is set as it would request an acknowledgment without sending payload. For this reason, the bit combination a_0b_1 is not allowed by the imaginary network protocol and must be prevented in the grammar of the cover carrier.

We assume that the bit mapping of step 5 and all earlier steps were already performed successfully. The designer will probably have selected none or a very unlikely message for the bit combination $a_0 b_0$ as it indicates an error each time it occurs and may not be stealthy for this reason.

G_{CP} is the regular grammar of the control protocol and G_{CC} is the regular grammar of the cover carrier. The grammars are defined as follows:

$$G_{CP} = (V, \sum, P, S) \tag{4.1}$$
$$V = \{S, A, B\} \tag{4.2}$$
$$\sum = \{a_0, a_1, b_0, b_1\} \tag{4.3}$$
$$P = \{S \rightarrow AB, \tag{4.4}$$
$$A \rightarrow a_1 | a_0, \tag{4.5}$$
$$B \rightarrow b_1 | b_0\} \tag{4.6}$$

$$G_{CC} = (V, \sum, P, S) \tag{4.7}$$
$$V = \{S, A, B\} \tag{4.8}$$
$$\sum = \{a_0, a_1, b_0, b_1\} \tag{4.9}$$
$$P = \{S \rightarrow Ab_0 | a_1 B, \tag{4.10}$$
$$A \rightarrow a_1 | a_0, \tag{4.11}$$
$$B \rightarrow b_1 | b_0\} \tag{4.12}$$

Using this setup, $L(G_{CP})$ comprises the words $a_0 b_0, a_0 b_1, a_1 b_0$, and $a_1 b_1$, while $L(G_{CC})$ comprises the words $a_0 b_0, a_1 b_0$, and $a_1 b_1$ (cf. Figure 4.9).

The word $a_0 b_1$ can only be produced in the control protocol. If the control protocol would reach a state in which the word $a_0 b_1$ is generated, that is, the bit combination in the cover carrier is set, it would break the rules of the used carrier (the state is marked gray in Figure 4.9). The language inclusion test thus fails and the protocol designer can modify the bit mapping (step 5), redesign the control protocol (step 3), or redesign the cover carrier (step 1).

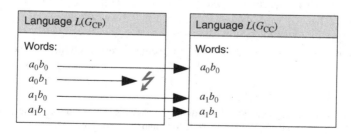

Figure 4.9. Produced words by the exemplary grammars G_{CP} and G_{CC}.

4.4 ADAPTIVE AND AUTONOMOUS COVERT CONTROL CHANNELS[4]

In addition to the development of covert channel–internal control protocols, other major ways to improve the resilience of covert communications are *autonomous* or *adaptive* covert channels.

A first approach for the development of an adaptive network covert channel is presented by Yarochkin et al. [26]. The authors developed an algorithm that is capable of switching the used application layer protocols of a covert channel. Additionally, the algorithm is adaptive, that is, it can select usable protocols and can monitor whether a protocol is blocked or not, even if a network configuration changes.

Yarochkin et al. support redundancy (resending the same content again over various channels) to counter payload transmission errors. The idea to utilize only protocols that are common within a given network to prevent IDS alerts is mentioned, but not discussed.

For the communication between two systems (called *agents*), the intersection of supported protocols between both hosts, excluding the set of blocked protocols, must be determined: $P_A \cap P_B \setminus P_{blocked}$. Therefore, the communication between two systems is split into two phases: the *network environment learning* phase (NEL phase) and the *communication* phase. Within the NEL phase, the systems try to determine usable protocols for the communication between them, while the communication phase starts after at least one usable protocol was determined in the NEL phase. The communication phase is responsible for transferring the hidden information. The NEL phase will continue for the whole lifetime of the covert communication processes to adapt the communication phase to changes in the network.

Within the NEL phase, the sender transmits a sequence of packets to the receiver in the hope that the receiver can identify the sequence using passive traffic monitoring. If the sequence is identified, the receiver acknowledges the successful data transmission.[5] Yarochkin et al. also calculate the round-trip time (RTT) for the response of packets of an *already used* protocol what is done to handle network problems (e.g., packet drops). In addition, a *survival score* is calculated for all already used protocols. The better the acknowledgment rate for a protocol, the higher the survival score of the protocol. To prevent those protocols, which were considered blocked, will stay unused after the blocking was administratively removed, a randomized protocol probing is carried out from time to time.

While Yarochkin et al. focus solely on network protocols, Wendzel and Keller in [2,3,27] introduce a finer-grained concept that works on the subcarrier level, that is, it allows

- that covert data are distributed over different subcarriers within the same network layer,

---

[4] This section is based on S. Wendzel, Novel Approaches for Network Covert Storage Channels, Ph.D. thesis, University of Hagen, 2013 and was slightly modified to fit into the context of this book.

[5] As shown by Wendzel, the NEL phase results in an unreliable link if an active warden is located on the path between two agents [27].

- that covert data are distributed over different network layers (e.g., transport layer and application layer can be used simultaneously),
- that different hiding methods for utilization of the same carrier are used depending on goals (e.g., maximum throughput or minimized packet count, cf. Section 4.4.1).

Li and He in 2011 presented an approach called autonomous covert channel [28], which is similar to Yarochkin's approach. Their idea to calculate survival values for covert channel transmissions is similar to the idea of Yarochkin et al. Like Yarochkin et al., Li and He do not describe the details of the acknowledgment channel that informs the sender about the success of transmissions, that is, the authors did not deal with the problem of active wardens (cf. [27]), nor is a covert channel–internal control protocol mentioned or described. The acknowledgment channel including the feedback information for the success of a transmission was left for future work by the authors [28].

Acosta and Medrano presented a blending covert method [29]. The method consists of three steps: a monitoring phase that analyzes received traffic passively (this idea is very similar to the previously discussed idea of Yarochkin et al., but takes into account payload fields and media streams, such as VoIP traffic, instead of protocol headers); a selection phase that is used to identify areas in network streams that could be used to embed hidden information into; and a transfer phase in which the actual embedding, transfer, and extraction of the hidden information is done. The third phase also includes a control protocol using synchronization information and a checksum. In [30], the authors additionally discuss the optimal placement of hidden information into network data and conclude that media streams can be considered as most suitable (*due to high update rates and high complexity* [30]).

Another independently developed approach was published in December 2012. Swinnen et al. proposed to monitor network traffic to adapt the traffic generation behavior of a covert channel to the traffic statistics of a given network [31]. Passive traffic monitoring is used to make daily traffic observations, since traffic patterns can change on a daily basis [31].

4.4.1 Optimized Post-NEL Communication

After the initial NEL phase is completed, carriers must be selected for the communication. Since different carriers can raise different attention, their usage rates can be configured by the covert channel sender. To minimize the raised attention, a PSCC can, as proposed by Wendzel and Keller [3], also minimize the required number of packets for a covert transaction or the overhead of a transaction. The optimization depends on the user's demands, especially on the use case. Therefore, [3] introduces the value q_i as the number of bits of the carrier C_i that is required to be transferred for one bit of s_i (the cover area):

$$q_i := \frac{\text{sizeof}(C_i)}{\text{sizeof}(s_i)}. \qquad (4.13)$$

Figure 4.10. The sender S transfers information to the receiver R via the covert channel proxies $Q_1 \cdots Q_n$. (Reproduced from [3] with permission of Springer.)

If linear optimization is applied, the optimal probabilities for the carriers can be calculated [3]. Therefore, it is useful to assign each carrier at least a small probability of occurrence to make forensic traffic analyses harder [3].

To optimize the throughput of the channel, the function f_1 can be maximized by adjusting the probabilities (p_i is the probability of carrier C_i):

$$f_1 = \sum_{i=1}^{n} p_i \cdot \text{sizeof}(s_i). \tag{4.14}$$

On the other hand, the function f_2 can be minimized to reduce the overhead required to transfer a hidden message [3]:

$$f_2 = \sum_{i=1}^{n} p_i \cdot q_i. \tag{4.15}$$

Control protocols can be used to exchange information about the optimization of covert channel traffic between two peers. Priority bits can be integrated into control protocol headers to indicate preferences (e.g., optimized throughput), similar to the quality of covertness introduced by Backs et al. [13].

4.4.2 Forwarding in Covert Channel Overlays

The minimization of packet count and overhead cannot be applied only to the direct communication between two peers but also to the forwarding of data over covert hops. Such hops must be aware of the covert channel and capable to forward the hidden information. Placing multiple hops between a covert channel sender and a receiver can improve the antitraceability of the communication and thus improve the safety of the communicating parties.

In this section, we assume that a control protocol takes care of exchanging meta-information about usable protocols (cf. Section 4.4) and that the performed routing paths are static. Due to the static routes, the optimized forwarding may not be optimal in the sense that no better option can be found for an overlay network.[6]

Figure 4.10 visualizes a proxy communication between the sender S and the receiver R using the covert channel proxies $Q_1 \cdots Q_n$ in a covert overlay network.

---

[6] Dynamic overlay routing for covert channels was explained in Chapter 4.4, too, but uses Quality of Covertness (QoC) optimization. So far, no implementation of a dynamic overlay routing for covert channels exists which minimizes packet count or overhead.

We assume that S and each forwarding instance Q_i choose one carrier to forward data to the next hop of the proxy chain. By applying the NEL phase, each host pair $(S, Q_1), (Q_1, Q_2), \cdots$, and (Q_n, R) can determine the available carriers they can use for communication in advance.

We assume that U_i is the set of usable carriers between element i and $i + 1$ of the chain and was determined in the NEL phase (S is element 0 and the receiver is element $n + 1$, while proxy Q_1 is element 1 and proxy Q_n is element n).

In the simplest case (i.e., no carriers are modified by an active warden), U_i is the intersection of the carriers $P()$ supported by element i and $i + 1$ of the chain, that is, $U_i = P(Q_i) \cap P(Q_{i+1})$. Let $s_{max}(i)$ be the maximal cover area space available per packet of all elements in U_i and let $q_{min}(i)$ be the minimal overhead of all elements in U_i. We can now apply the optimization of [3] for the forwarding in the proxy chain as follows if a packet from element Q_i is received at element Q_{i+1} and must be forwarded to Q_{i+2} (Q_0 would be the sender and Q_{n+1} the receiver).

Optimize covert transaction for minimal packet count on hop Q_{i+1}:

1. If $s_{max}(i) = s_{max}(i + 1)$, then forward the payload directly to Q_{i+2}.
2. If the received packet is the last packet of the transaction (e.g., indicated by a flag in the control protocol [8]), forward the data to Q_{i+2} using the carrier with $s_{max}(i + 1)$.
3. Otherwise, forward as many complete covert data packets using the carrier that provides $s_{max}(i + 1)$ as feasible. As pointed out in [3], bursts in case of $s_{max}(i + 1) \ll s_{max}(i)$ can be provided by forcing a maximum packet frequency via the leaky bucket method. If there are no data left to be forwarded, wait for time t for new data to arrive and afterward forward it using the carrier that provides $s_{max}(i + 1)$. This step must be repeated until no new data arrive for the time t or the last packet of the transaction was received.

To prevent that only the optimal carrier is used, that is, to also enable other carriers, each carrier C_i can be used with at least a small occurrence rate ($p_i > 0$).

Optimize covert transaction for minimal overhead on hop Q_{i+1}: To minimize the overhead of the covert channel transaction, the previous algorithm has to use $q_{min}(i)$ instead of $s_{max}(i)$ and $q_{min}(i + 1)$ instead of $s_{max}(i + 1)$:

1. If the received message size $sizeof(m)$ is of the same size as space is provided by $q_{min}(i + 1)$, then directly forward the data using the carrier that provides $q_{min}(i + 1)$.
2. If the transaction ends, send all remaining information using the carrier that provides $q_{min}(i + 1)$ as long as enough data remain that have the size of the space

of the carrier that provides $q_{min}(i + 1)$. Afterward use the carrier $C_j \in U_{i+1}$ for which the overhead of the remaining k bits is minimal.[7]

3. Otherwise, forward as many full covert data packets using the carrier that provides $q_{min}(i + 1)$ as possible. Therefore, wait for time t for new traffic to arrive. If no new data arrive, use the carrier that provides the smallest overhead for the remaining k bits to forward the data. This step must be repeated until no new data arrive for the time t or if the last packet of the transaction was received.

4.4.3 Influence of Steganographic Attributes

As adaptive control protocols can configure the steganographic channel (e.g., using (sub)carrier selection) and enhance the channel's capabilities (e.g., with dynamic overlay routing or reliability), they influence the three steganographic features shown in Figure 2.7 as well as the steganographic cost (cf. Section 2.3). In other words, steganographic control protocols influence and can optimize

- the detectability of the channel (by selecting a (sub)carrier, hiding method, and bandwidth),
- the bandwidth of the channel (by agreeing with steganographic peers on a given bandwidth and (sub)carrier, on which the bandwidth depends),
- the robustness of the channel (by selecting an appropriate hiding method and (sub)carrier), and
- the steganographic cost of the channel (by selecting a (sub)carrier, a hiding method, and the bandwidth of the channel).

4.5 TECHNIQUES FOR TIMING METHODS

So far, the techniques discussed in this chapter were applied only to storage channels but not to timing channels. In the case of timing channels, the placement of control protocol headers in a "storage" form is not feasible as the carrier provides no suitable storage space (i.e., bits in a PDU).

Instead, timing channels can utilize coding techniques to signal hidden information so that a reliable data transfer can be achieved. Knowing a timing channel's capacity, the sender and receiver can define a robust signaling with a lower throughput. For instance, they can agree on a reduced number of code symbols that they can easily distinguish.

However, to deal with noise, it is crucial to apply techniques that allow the detection of transmission errors, for example, by using checksums, error correcting codes, or (re)synchronization sequences. It is also feasible to combine storage and timing channels

---

[7] It is thinkable that $q_{min}(i + 1)$ is the only optimal if all provided space of the cover carrier is used, but a carrier providing fewer bits could be more suitable for the transfer of the remaining k bits.

to hybrid channels (Section 3.3). Then, a more robust storage channel can be used to ensure reliable communication over a timing channel.

A covert timing channel that provides a reliable communication among multiple TCP flows was presented by Lua et al. in [32]. Another timing channel that utilizes the permutations of multiple distinguishable, sequential network objects (e.g., HTTP requests or FTP commands) was presented by Swinnen et al. [33] and an implementation called *CoCo* by Houmansadr and Borisov creates a reliable timing channel based on the modulation of inter packet gaps [34].

4.6 ATTACKS ON CONTROL PROTOCOLS

Like many other network protocols, control protocols are vulnerable to a number of attacks. For instance, an attacker can perform man-in-the-middle attacks to redirect control protocol messages or impersonate another steganographer to spoof commands, for example, to terminate a connection or to switch another hiding method. A fundamental difference between countermeasures for nonhidden network protocols and countermeasures for control protocols is that an attacker may not know whether a control protocol is present or how it is embedded into the carrier. We discuss countermeasures on control protocols in detail in Section 8.1.2.

4.7 OPEN RESEARCH CHALLENGES FOR CONTROL PROTOCOLS

As this chapter demonstrated, the spectrum of control protocol-related topics is broad. A number of known challenges in control protocol research remain, which are summarized in this section. These research topics are based on the ones already proposed in [4] and were modified and extended for this chapter.

4.7.1 Multilayer Control Protocols

None of the known steganographic control protocols provides a clear separation of layers as, for instance, can be found in the case of TCP/IP protocols. In TCP/IP, one protocol data unit of a lower layer encapsulates a protocol data unit of a higher layer—for instance, an Ethernet frame encapsulates IPv4, IPv4 encapsulates UDP, and UDP encapsulates DNS. The research questions that arise in this context are as follows: Is it useful to split a steganographic control protocol among different layers? If so, how should it be designed in an optimal manner to be space-efficient, given the limited bandwidth? Research on multilayer control protocol design should also take existing techniques of protocol compression and space optimization into account, such as *status updates*.

4.7.2 Protocol Translation for Control Protocols

Given the currently rather unlikely coexistence of multiple covert overlay networks, these overlay networks can utilize different steganographic control protocols. An interconnectivity of larger, separated but incompatible, covert overlay networks would be feasible if protocol translation mechanisms are provided for control protocols. Challenging aspects in this regard are as follows:

(i) Different control protocols provide different feature sets, that is, a feature provided by control protocol P_a may not be provided by P_b and can thus not be handled in environments using P_b. Alternatively, some features could be provided in a different manner. An example scenario would be that two protocols P_a and P_b implement a reliability mechanism based on different reliability algorithms, each probably utilizing a different size for the sequence number.

(ii) Control protocols can provide overlay addressing schemes. To this end, they do not solely utilize IPv4 or IPv6 addresses for their communications, but shorter, internal addresses. These addresses can be redundant or incompatible to addresses used in other covert overlay networks with other protocols. Such a problem may be reduced by network address translation (NAT) for steganographic communications. It is also thinkable to tunnel one covert communication within another to bypass incompatible network segments similarly as IPv6 can be tunneled through IPv4 subnets (similar to [20]).

4.7.3 Handling of Fake-Input

Some control protocols already provide simple means to detect fake control protocol messages as well as the accidental interpretation of noncovert network traffic as covert traffic. Ping tunnel includes a magic number set to a given value and all ICMP Echo packets that do not provide the correct four bytes at the beginning of the payload will be ignored. A magic number is, however, no satisfying solution to prevent intentional attacks using fake control protocol headers, as described in Section 8.1.2. Only the hybrid approach (Section 4.1.3.5) provides authentication and integrity protection for covert data transfers.

4.8 SUMMARY

Control protocols enrich a steganographic communication with reliability, clear communication procedures, optimization (especially for stealthiness), and adaptiveness (i.e., adjustability to changes in a network environment). The known control protocols moreover enable dynamic routing for covert overlays and carrier switching—just to mention a few features. Using these features, control protocols influence and can optimize the detectability, bandwidth, robustness, and cost of a steganographic channel.

Control protocols are also called *micro-protocols* and are either embedded into the steganogram together with the covert payload or are represented by timing information. The "size" of a control protocol must be kept small in order to leave enough bandwidth to transfer covert payload and a measure called *features per header bit* is used to indicate the space-efficiency of a control protocol's design. Two protocol engineering means for storage channels exist, which optimize the size and the embedding of a control protocol in a way that it blends with the carrier.

This chapter discussed and compared existing control protocols and the developments in the area of adaptive control channels. Adaptive control channels can bypass administrative changes or restrictions in networks by observing the network environment in which they operate. As we will show in Section 8.1.2, attacks on control protocols are also feasible.

Further research is required to provide an outlook on the benefits and feasibility of a multilayer architecture for control protocols. Another area left for future research is the interconnection of isolated covert overlay networks operating with different control protocols. Further research may also introduce first engineering means for control protocols in timing channels.

REFERENCES

1. N. Nounou and Y. Yemini. Development tools for communication protocols. Technical Report, Computer Science Department, Columbia University, New York, 1985.

2. S. Wendzel and J. Keller. Systematic engineering of control protocols for covert channels. In *Proceedings of the 13th Conference on Communications and Multimedia Security (CMS'12)*, Vol. 7394 of *Lecture Notes in Computer Science*, pp. 131–144. Springer, 2012.

3. S. Wendzel and J. Keller. Low-attention forwarding for mobile network covert channels. In *Proceedings of the 12th Conference on Communications and Multimedia Security (CMS'11)*, Vol. 7025 of *Lecture Notes in Computer Science*, pp. 122–133. Springer, October 2011.

4. S. Wendzel and J. Keller. Hidden and under control: a survey and outlook on covert channel–internal control protocols. *Annals of Telecommunications*, 69(7):417–430, 2014.

5. S. Wendzel, B. Kahler, and T. Rist. Covert channels and their prevention in building automation protocols: a prototype exemplified using BACnet. In *Proceedings of the Workshop on Security of Systems and Software Resiliency (3SL'12)*, pp. 731–736. IEEE, September 2012.

6. D. Stødle. Ping tunnel: for those times when everything else is blocked, September 2011. http://www.cs.uit.no/ daniels/PingTunnel/.

7. R. deGraaf, J. Aycock, and M. J., Jr. Improved port knocking with strong authentication. In *Proceedings of the 21st Annual Computer Security Applications Conference (ACSAC'05)*, pp. 451–462. IEEE Computer Society, 2005.

8. B. Ray and S. Mishra. A protocol for building secure and reliable covert channel. In *Proceedings of the 6th Annual Conference on Privacy, Security and Trust (PST'08)*, pp. 246–253. IEEE, 2008.

9. K. R. Fall and W. R. Stevens. *TCP/IP Illustrated, Volume 1: The Protocols, 2nd revised edition*. Addison-Wesley Professional Computing Series. Addison-Wesley, 2011.

10. Z. Trabelsi and I. Jawhar. Covert file transfer protocol based on the IP record route option. *Journal of Information Assurance and Security*, 5(1):64–73, 2010.

11. W. Mazurczyk and Z. Kotulski. New security and control protocol for VoIP based on steganography and digital watermarking. *Annales UMCS, Informatica*, 5:417–426, 2006.

12. K. Szczypiorski, I. Margasinski, et al. TrustMAS: trusted communication platform for multi-agent systems. In *Proceedings of the OTM 2008*, Vol. 5332 of *Lecture Notes in Computer Science*, pp. 1019–1035. Springer, 2008.

13. P. Backs, S. Wendzel, and J. Keller. Dynamic routing in covert channel overlays based on control protocols. In *Proceedings of the International Workshop on Information Security, Theory and Practice (ISTP '12)*, pp. 32–39. IEEE, 2012.

14. B. Jankowski, W. Mazurczyk, and K. Szczypiorski. Information hiding using improper frame padding. In *14th International Telecommunications Network Strategy and Planning Symposium (NETWORKS)*, pp. 1–6. IEEE, 2010.

15. S. Wendzel. Protocol hopping covert channels. November 2007. http://www.wendzel.de/dr.org/files/Papers/protocolhopping.txt.

16. W. Fraczek, W. Mazurczyk, and K. Szczypiorski. How hidden can be even more hidden? In *Proceedings of the First International Workshop on Digital Forensics (IWDF '11)*, November 2011.

17. W. Mazurczyk, S. Wendzel, I. A. Villares, and K. Szczypiorski. On importance of steganographic cost for network steganography. *International Journal of Security and Communication Networks*, 2014. DOI: 10.1002/sec.1085.

18. daemon9. LOKI2 (the implementation), September 1997. http://gray-world.net/papers/projectloki2.txt.

19. S. Wendzel. Protocol channels as a new design alternative of covert channels. Technical Report CORR, vol. abs/0809.1949, Kempten University of Applied Sciences, September 2008. http://arxiv.org/abs/0809.1949.

20. W. Fraczek, W. Mazurczyk, and K. Szczypiorski. Multilevel steganography: improving hidden communication in networks. *Journal of Universal Computer Science*, 18(14):1967–1986, 2012.

21. S. Wendzel and J. Keller. Preventing protocol switching covert channels. *International Journal on Advances in Security*, 5(3-4):81–93, 2012.

22. S. Wendzel and S. Zander. Detecting protocol switching covert channels. In *Proceedings of the 37th IEEE Conference on Local Computer Networks (LCN)*, pp. 280–283, 2012.

23. G. J. Holzmann. *Design and Validation of Computer Protocols*. Prentice Hall, 1991.

24. V. Jacobson. Compressing TCP/IP headers for low-speed serial links (RFC 1144), 1990. http://www.rfc-editor.org/rfc/rfc1144.txt.

25. V. Gorodetski and I. Kotenko. Attacks against computer network: formal grammar-based framework and simulation tool. In *Recent Advances in Intrusion Detection*, Vol. 2516 of *Lecture Notes in Computer Science*, pp. 219–238. Springer, Berlin, 2002.

26. F. V. Yarochkin, S.-Y. Dai, C.-H. Lin, et al. Towards adaptive covert communication system. In *Proceedings of the 14th IEEE Pacific Rim International Symposium on Dependable Computing (PRDC'08)*, pp. 153–159. IEEE Computer Society, 2008.

27. S. Wendzel. The problem of traffic normalization within a covert channel's network environment learning phase. In *Proceedings of the Sicherheit'12*, Vol. P-195 of *Lecture Notes in Informatics*, pp. 149–161, 2012.

28. W. Li and G. He. Towards a protocol for autonomic covert communication. In *Proceedings of the 8th International Conference on Autonomic and Trusted Computing (ATC'11)*, pp. 106–117. Springer, 2011.

29. J. Acosta and J. Medrano. Using a novel blending method over multiple network connections for secure communication. In *Proceedings of the Military Communications Conference 2011 (MILCOM '11)*, pp. 1460–1465, 2011.

30. J. Acosta and J. Medrano. NBCS: secure communication via distributed covert channels in active network traffic. Unpublished; available online under http://www.researchgate.net/publication/228925608_NBCS_Secure_communication_via_distributed_covert_channels_in_active_network_traffic.

31. A. Swinnen, R. Strackx, P. Philippaerts, et al. Protoleaks: a reliable and protocol-independent network covert channel. In *Proceedings of the 8th International Conference on Information Systems Security (ICISS'12)*, Vol. 7671 of *Lecture Notes in Computer Science*, pp. 119–133. Springer, 2012.

32. X. Luo, E. Chan, and R. Chang. Cloak: a ten-fold way for reliable covert communications. In *Computer Security: ESORICS 2007*, Vol. 4734 of *Lecture Notes in Computer Science*, pp. 283–298. Springer, 2007.

33. A. Swinnen, R. Strackx, P. Philippaerts, and F. Piessens. Protoleaks: a reliable and protocol-independent network covert channel. *Information Systems Security*, 7671:119–133, 2012.

34. A. Houmansadr and N. Borisov. CoCo: coding-based covert timing channels for network flows. In *Proceedings of the 13th International Conference on Information Hiding*, pp. 314–328. Springer, 2011.

5

TRAFFIC TYPE OBFUSCATION

If the only tool you have is a hammer, you tend to see every problem as a nail.

—Abraham Maslow

Traffic type obfuscation (TTO) is hiding the type of network traffic, that is, the underlying network protocol, exchanged between two (or multiple) endpoints. Suppose that two Internet endpoints Alice and Bob exchange Internet traffic using an Internet protocol P_1. Alice and Bob may choose to use traffic type obfuscation so that a third party, Eve, is not able to identify P_1 as the protocol used by Alice and Bob for their communication.

In order to be effective, a typical traffic type obfuscation scheme may need to modify one or several of the following traffic features:

- content of traffic, such as IP packet content;
- patterns of traffic, such as IP packet timings and sizes;
- protocol behavior, such as reaction to network perturbations.

Information Hiding in Communication Networks: Fundamentals, Mechanisms, Applications, and Countermeasures,
First Edition. Wojciech Mazurczyk, Steffen Wendzel, Sebastian Zander, Amir Houmansadr, and Krzysztof Szczypiorski.
© 2016 by The Institute of Electrical and Electronics Engineers, Inc. Published 2016 by John Wiley & Sons, Inc.

5.1 PRELIMINARIES

5.1.1 Applications

Traffic type obfuscation techniques are utilized for two main applications.

- *Blocking resistance:* Traffic type obfuscation is used as a countermeasure in application scenarios where the use of particular types of network protocols is prohibited. For instance, some Internet ISPs disallow the access to P2P file sharing protocols such as BitTorrent,[1] and some enterprise networks block the use of Skype.[2] To enforce such blocking, ISPs use different networking mechanisms [1–3] to identify (and then block) the network traffic running the disallowed protocols. As another example, repressive regimes prohibit the use of systems that circumvent Internet censorship, for example, Tor [4]. Similar networking mechanisms are used by censorship technologies to detect the traffic generated by prohibited circumvention software [5–8].
- *Privacy protection:* Traffic type obfuscation is alternatively used to augment Internet users' privacy, for example, by manipulating traffic patterns to conceal the intent of communication. For instance, previous research [9–11] shows that an adversary can learn sensitive information about the Internet browsing activities of users (e.g., the websites they visit) by statistically analyzing their encrypted web traffic.

5.1.2 Comparison with Network Steganography

As noted in Chapter 1, traffic type obfuscation shares similarities with network steganography, the other category of information hiding in communication networks introduced in Chapter 3. Traffic type obfuscation resembles network steganography in that both aim to hide the *existence of a covert communication*. On the other hand, while the main objective of network steganography is to communicate the highest amount of covert information without being detected, the main objective of traffic type obfuscation is to conceal the *type* of network traffic. As a result, features such as bandwidth, robustness, and cost (which are used to evaluate network steganography) are out of scope for type obfuscation. Instead, a type obfuscation scheme should be *unobservable*: third-parties should not be able to (1) detect that the traffic has been obfuscated, and, in some applications, (2) disclose the obfuscated traffic type.

---

[1] www.bittorrent.com/.

[2] www.skype.com/.

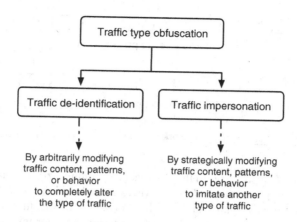

Figure 5.1. Classes of traffic type obfuscation based on the objective.

5.2 CLASSIFICATION BASED ON THE OBJECTIVE

The existing work on traffic type obfuscation can be broadly classified into two categories, as shown in Figure 5.1: *traffic de-identification* and *traffic impersonation*.

5.2.1 Traffic De-identification

Traffic de-identification manipulates network traffic so that the underlying network protocol is concealed. For instance, network packet headers are encrypted to conceal protocol identifier contents, thus obfuscating the underlying network protocol. In the following, some example approaches toward traffic de-identification are presented.

5.2.1.1 Padding Encrypted Packets. While encryption hides packet contents, traffic patterns, such as packet timings and packet sizes, can potentially reveal critical information about the content being communicated. For instance, previous work has shown that analyzing the patterns (e.g., packet sizes) of encrypted web traffic can give away the identities of the websites visited by users [11–14].

To counter this, several proposals have suggested to pad encrypted packets to conceal the actual lengths of plaintext packets. In this approach, already deployed in some implementations of the transport layer protocol (TLS), for example, GnuTLS,[3] random numbers of extra bytes are added at the end of packets before they are encrypted (see Figure 5.2). The SSH, TLS, and IPSec protocols (which are "prevalent" protocols for encrypting network traffic) allow up to 255 bytes of padding in their design. Previous work [9] discusses various approaches to pad the encrypted traffic of these protocols:

---

[3] http://www.gnu.org/software/gnutls/.

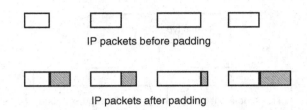

Figure 5.2. Padding network packets to de-identify packet sizes.

1. *Session random padding:* In this approach, a padding size, p, is randomly chosen from the set $\{0, 8, 16, ..., M - \ell\}$, where ℓ is the packet size before padding and M is the maximum transmission unit (MTU), that is, the size of the largest protocol data unit defined in the protocol's standard. For instance, the MTU is 1500 bytes for IEEE 802.3. Every packet in the target session is padded with a dummy plaintext of size p.
2. *Packet random padding:* In contrast to the session padding approach, every single packet is padded with a dummy content of a randomly chosen size, independently chosen across all of the packets in a session.
3. *Linear padding:* In this approach, packets are padded so that all packet sizes are increased to the nearest multiple of 128, or the MTU, whichever is smaller.
4. *Exponential padding:* All packets are padded to the nearest power of 2, or the MTU, whichever is smaller.
5. *Mice-Elephants padding:* The packets smaller than 128 bytes are padded to 128 bytes; all the other packets are padded to the MTU.
6. *Pad to MTU:* All packets are padded to the MTU size.

Note that some of these padding approaches are already deployed by some implementations of secure network protocols, such as GnuTLS.

5.2.1.2   HTTPOS. As noted earlier, traffic analysis can infer sensitive information from the communication patterns of encrypted traffic. In the context of web traffic, previous research [10–14] showed that traffic analysis can infer sensitive information about the browsing activities of a user by analyzing his/her encrypted web traffic, for example, HTTPS traffic (in this context, de-identification aims at masking the identities of HTTP destinations, not the HTTP protocol itself). Previous work has taken several directions to foil this, including the server-side techniques [15,16] that require modification of web entities, such as servers, browsers, and web objects, to deploy protocol de-identification. HTTPOS [17] is an alternative approach for safeguarding encrypted HTTP traffic against identification. HTTPOS deploys a browser-side approach; that is, obfuscation mechanisms are implemented at the client machine without the need to modify any web entity.

HTTPOS obfuscates traffic by modifying four fundamental features of TCP and HTTP protocols: packet sizes, timing of packets, sizes of web objects, and flow sizes.

Note that since HTTPOS works at the client side, it cannot directly modify such traffic features on the traffic from web servers. To be able to do so, HTTPOS exploits a number of basic features in the HTTP protocol, for example, HTTP pipelining and HTTP range, and TCP protocol, for example, maximum segment size negotiation and advertised window.

5.2.2 Traffic Impersonation

As introduced earlier, the aim of traffic impersonation is to not only hide the underlying network protocol, but also pretend to be using another type of protocol, that is, a target protocol. For instance, a BitTorrent client may modify her traffic to look like it is carrying HTTP traffic.

5.2.2.1 SkypeMorph. Tor [4] is a popular system used by Internet users living under repressive regimes to bypass Internet censorship. Unfortunately, today's modern networking technology allows state-level censors, that is, totalitarian governments, to identify and block the use of Tor's network protocol. This is done in different ways, such as statistical analysis of Tor traffic patterns [18] and looking for Tor protocol-specific identifiers [19, Slide 38]. To be usable, it is of paramount importance for Tor to deploy mechanisms that conceal the use of the Tor protocol at the network layer. *Pluggable transports* [20] are Tor's recent initiative toward protecting the identity of Tor traffic (and Tor parties) from the censors.

SkypeMorph [21] is a pluggable transport for Tor. It intends to conceal Tor's traffic by making the traffic between a Tor client and a Tor server look like a Skype video call. That is, a Tor user running SkypeMorph software *mimics* Skype's network protocol.

Figure 5.3 illustrates SkypeMorph's architecture. The following summarizes how SkypeMorph works to mimic Skype:

1. The SkypeMorph server creates a legitimate Skype ID, and shares the ID with the SkypeMorph client. The server, then, logs in to Skype, and listens for incoming Skype calls.

2. The SkypeMorph client similarly creates a legitimate Skype ID.

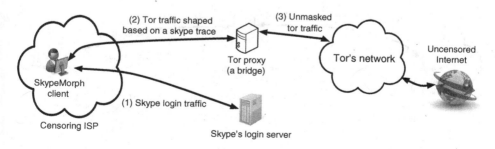

Figure 5.3. The main architecture of SkypeMorph [21].

3. The client checks to see whether the server's Skype ID is online. If so, the client uses Skype's text messaging platform to share some cryptographic keys with the server. The server and client send a few other Skype text messages to finish this initial handshake, which results in sharing cryptographic keys and other secret information needed for the operation of SkypeMorph.

4. The SkypeMorph client starts a Skype video call with the server's Skype ID. To do so, it uses Skype's API to send a ''ringing'' message to the server for a random amount of time.

5. Both the client and the server exit Skype's API. They establish a Tor connection; however, using the cryptographic keys derived in previous steps, they encrypt Tor packets to hide Tor's protocol. Additionally, they *shape* the Tor traffic; that is, they use network traces of Skype traffic to mimic the timings and packet sizes of a previously established Skype conversation.

Analysis of SkypeMorph [21] demonstrates that the traffic patterns (e.g., packet size distribution and interpacket delay distribution) of a Tor connection over SkpeMorph resemble those of Skype traffic, not Tor.

5.2.2.2 FreeWave. FreeWave [22] is a blocking-resistant system that aims at hiding the use of a disallowed network protocol. For instance, it can be deployed by Tor entry points as a Tor pluggable transport [20] in order to disguise Tor traffic, or can be used by BitTorrent clients to obfuscate their file sharing traffic. Similar to SkypeMorph, introduced above, FreeWave mimics the popular Voice-over-IP (VoIP) network protocol of Skype. On the other hand, contrary to SkypeMorph that shapes Tor traffic to look like Skype, FreeWave runs the actual Skype protocol for the whole duration of a connection, and encodes the source traffic (e.g., Tor traffic) into the acoustic signals carried over the Skype protocol. Figure 5.4 illustrates the architecture of FreeWave.

The following are the main components of FreeWave client and server software:

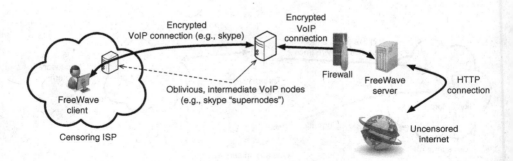

Figure 5.4. The main components of FreeWave [22].

1. *VoIP client:* VoIP client is a Voice-over-IP client software, such as Skype software, that allows VoIP users to connect to one (or more) specific VoIP service(s).

2. *Virtual sound card (VSC):* A virtual sound card (VSC) is a software application that uses a physical sound card installed on a machine to generate one (or more) isolated, virtual sound card interfaces on that machine. A virtual sound card interface can be used by any application running on the host machine exactly the same way a physical sound card is utilized. Also, the audio captured or played by a virtual sound card does not interfere with that of other physical/virtual sound interfaces installed on the same machine. FreeWave uses VSCs to isolate the audio signals generated by its software from the audio belonging to other applications.

3. *MoDem:* FreeWave client and server software use a modulator/demodulator (MoDem) application that translates network traffic into acoustic signals and vice versa. This allows FreeWave to tunnel the network traffic of its clients over VoIP connections by modulating them into acoustic signals. Houmansadr et al. [22] use communication codes to design a reliable MoDem that withstands noise from the network and Skype codec.

4. *Proxy:* The FreeWave server uses an ordinary network proxy application that proxies the network traffic of FreeWave clients, received over VoIP connections, to their final Internet destinations. Two popular choices for FreeWave's proxy are the HTTP proxy [23] and the SOCKS proxy [24].

FREEWAVE'S CLIENT DESIGN. The FreeWave client software consists of three main components described above: a VoIP client application, a virtual sound card, and the MoDem software. Figure 5.5 shows the block diagram of the FreeWave client design. The MoDem transforms the data of the network connections sent by the web browser into acoustic signals and sends them over to the VSC component. The FreeWave MoDem also listens on the VSC sound card to receive specially formatted acoustic signals that carry modulated Internet traffic; MoDem extracts the modulated Internet traffic from such acoustic signals and sends them to the web browser. In a sense, the client web browser uses the MoDem component as a network proxy; that is, the listening port of MoDem is entered in the HTTP/SOCKS proxy settings of the browser.

The VSC sound card acts as a bridge between MoDem and the VoIP client component; that is, it transfers audio signals between them. More specifically, the VoIP client is set up to use the VSC sound card as its "speaker" and "microphone" devices (VoIP

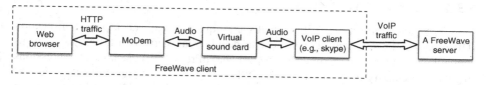

Figure 5.5. The main components of FreeWave [22] client.

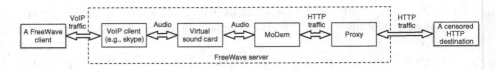

Figure 5.6. The main components of FreeWave [22] server.

applications allow a user to select physical/virtual sound cards). This allows MoDem and the VoIP client to exchange audio signals that contain the modulated network traffic, isolated from the audio generated/recorded by other applications on the client machine.

For the FreeWave client to connect to a particular FreeWave server, it requires to know the VoIP ID belonging to that server. Every time the user starts up the FreeWave client application on his/her machine, the VoIP application of FreeWave client initiates an audio/video VoIP call to the known VoIP ID of the FreeWave server.

FREEWAVE'S SERVER DESIGN. Figure 5.6 shows the design of FreeWave server, which consists of four main elements. FreeWave server uses a VoIP client application to communicate with its clients through VoIP connections. A FreeWave server chooses one or more VoIP IDs, which are provided to its clients, for example, through public advertisement.

The VoIP client of the FreeWave server uses one (or more) virtual sound card(s) as its "speaker" and "microphone" devices. The VSCs are also used by the MoDem component, which transforms network traffic into acoustic signals and vice versa. More specifically, MoDem extracts the Internet traffic modulated by FreeWave clients into audio signals from the incoming VoIP connections and forwards them to the last element of the FreeWave server, FreeWave proxy. MoDem also modulates the Internet traffic received from the proxy component into acoustic signals and sends them to the VoIP client software through the VSC interface.

5.3 CLASSIFICATION BASED ON THE IMPLEMENTATION DOMAIN

Traffic type obfuscation techniques are classified into several categories based on the domain they use to deploy obfuscation mechanisms, as shown in Figure 5.7.

5.3.1 Content-Based Schemes

Modern state-of-the-art network firewalls are equipped with deep-packet inspection (DPI) technologies, which allows them to deeply scan network traffic. Many open-source DPI systems (e.g., Bro [25], Yaf [26], and nProbe [27]) use regular expressions (regexes) to classify network protocols. DPIs look for content signatures that identify a particular network protocol; hence, they can be used to identify and block specific network protocols, for example, Skype and Tor protocols.

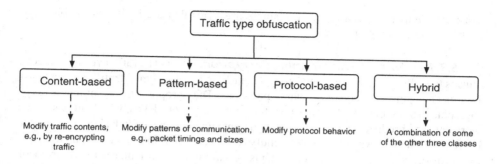

Figure 5.7. Classes of traffic type obfuscation based on the implementation domain.

Content-based traffic type obfuscation schemes aim at defeating DPI technologies by manipulating packet contents (e.g., using encryption) in order to remove content signatures that reveal the use of a particular network protocol. In the following, we introduce several content-based protocol obfuscators.

5.3.1.1 Tor Content Obfuscation.
In order to improve its undetectability against competent state-level censors, the Tor project has initiated the pluggable transports project. Pluggable transports aim at hiding the very existence of the Tor protocol by deploying various protocol obfuscation mechanisms, including content-based, pattern-based, and protocol-based techniques.

Obfsproxy [28] is the first Tor pluggable transport, which is a content-based obfuscation mechanism. Figure 5.8 shows the main block diagram of Obfsproxy. Obfsproxy works by re-encrypting Tor packets in order to conceal Tor-specific content signatures in Tor packets.

Obfsproxy is the basis of many subsequent pluggable transport designs, including SkypeMorph (Section 5.2.2.1), StegoTorus (Section 5.3.4.1), and FTE (Section 5.3.1.2). As another content-based example for pluggable transport, Dust [29] aims at defeating protocol fingerprinting through deep-packet inspection by transforming the contents of

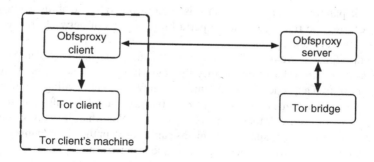

Figure 5.8. The main architecture of Obfsproxy.

Tor traffic into encrypted or random single-use bytes that are indistinguishable from each other and random packets.

5.3.1.2 Format-Transforming Encryption.

Dyer et al. [30] investigate mechanisms that allow an attacker to enforce content-based protocol misidentification against DPI mechanisms that rely on regular expressions. They propose the use of a new cryptographic primitive, named format-transforming encryption (FTE), which allows a network entity to produce encrypted content guaranteeing that they match a specific regular expression. In other words, a network entity (e.g., a censored Internet user) can use FTE to transform the regexes of packet contents pertaining to a particular network protocol (the source protocol) into the regexes of another protocol (the target protocol).

To be more specific, FTE is an encryption scheme that takes a regular expression as an input, and encrypts an arbitrary plaintext so that the generated ciphertext matches the input regular expression. Dyer et al. use FTE as a component within the TLS record layer to handle streams of messages from arbitrary source protocols.

FTE, which is inspired by the format-preserving encryption (FPE) [31], works in the following steps:

1. The plaintext message M is encrypted using a standard authenticated encryption scheme, resulting in an intermediate ciphertext Y.
2. Y is treated as an integer in $\mathbb{Z}_S$, which is the set of integers from 0 to the size of the language minus one. S is the set of expressions in the input language.
3. Y is encoded using function $unrank : \mathbb{Z}_S \to S$, generating the final ciphertext C. To decrypt, there needs to exist a reverse function for $unrank$, that is, $rank : S \to \mathbb{Z}_S$, that is efficiently computable.

Dyer et al. demonstrate that FTE can make several real-world DPI systems misidentify network protocols at the cost of small bandwidth overhead.

5.3.2 Pattern-Based Schemes

Previous work [9,16,32–34] demonstrates the feasibility of protocol classification by analyzing traffic patterns, that is, traffic analysis. Pattern-based traffic obfuscation schemes aim at concealing the type of traffic by perturbing patterns of network traffic, such as packet timings and packet sizes.

In Section 5.2.1.1, we described per-packet padding mechanisms that aim at modifying packet sizes in order to de-identify the underlying network protocol. Wright et al. [16] propose two approaches that augment such per-packet padding mechanisms by not only concealing the type of the underlying network protocol, but also mimicking that of another network protocol, that is, a target protocol. Note that, instead of mimicking patterns of another protocol, such mechanisms can be used in the same manner to mimic another instance of the same protocol. For example, a pattern-based obfuscation scheme can be used to disguise HTTP traffic with website A as HTTP traffic with another website, B.

5.3.2.1 Direct Target Sampling.

In this approach, network packets are padded with arbitrary content (e.g., random bits) in order to change the sizes of network packets whose pattern may potentially reveal the underlying protocol. The following is how the DTS protocol works. Consider a source protocol A and a target protocol B (or, alternatively, a source and target instance of the same protocol). Also, assume D_A and D_B as the respective probability distributions of their packet sizes. The direct target sampling (DTS) scheme works as follows:

1. For any packet of size i from the source protocol, the DTS scheme derives a packet size, i', from D_B at random.
2. If $i' \geq i$, DTS pads the input packet from A to size i', and sends the packet.
3. If $i' < i$, DTS sends the first i' bytes of the input packet as a single network packet. DTS, then, goes to step 1; that is, it continues sampling from D_B until all bytes of the original input packet are sent.

5.3.2.2 Traffic Morphing.

Traffic morphing works in a similar manner to direct target sampling, except it uses a different sampling mechanism. Contrary to direct target sampling, which directly samples from the target protocol (D_B), traffic morphing uses convex optimization methods to sample. More specifically, traffic morphing uses convex optimization to generate a *morphing matrix* that makes source traffic look like the target protocol while minimizing the overhead, for example, the number of packets sent.

Consider the source packet sizes as a vector of size n, $S_A = [x_1, x_2, ..., x_n]^T$ (T is the transpose operation). The output packet sizes are derived as

$$S_B = A \cdot S_A, \tag{5.1}$$

where $S_B = [y_1, y_2, ..., y_n]^T$. A is the morphing matrix of size $n \times n$, that is, $A = [a_{ij}]$ ($1 \leq i, j \leq n$). To morph A to B, each column of the morphing matrix A should be a valid probability mass function over the n packet sizes. For any input packet with size s_j, the morphing algorithm looks at the jth column of A and samples a target size s_i using the probability distribution of that column. Similar to the case of DTS, if the derived target size is larger than the input size, the input packet is padded and sent. Otherwise, the input packet is split into multiple packets recursively running the morphing algorithm.

The morphing matrix is derived using convex optimization so that it minimizes a cost function, $f_0(A)$, subject to multiple constraints. The constraints in this case are

$$\sum_{j=1}^{n} a_{ij} x_j = y_i, \qquad \forall i \in [1, n], \tag{5.2}$$

$$\sum_{i=1}^{n} a_{ij} = 1, \qquad \forall j \in [1, n], \tag{5.3}$$

$$a_{ij} \geq 0, \qquad \forall i, j \in [1, n]. \tag{5.4}$$

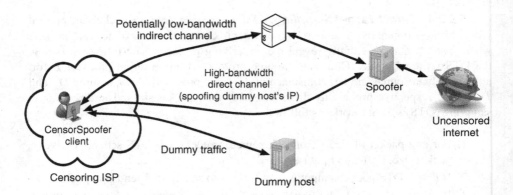

Figure 5.9. The main architecture of CensorSpoofer [36].

5.3.3 Protocol-Based Schemes

The third class of protocol type obfuscation includes protocol-based schemes. These techniques [21,35,36] modify the protocol behavior such as the handshaking mechanism and subprotocol dependencies in order to obfuscate protocol types. In the following, two protocol-based obfuscation schemes are described.

5.3.3.1 CensorSpoofer. CensorSpoofer is a system for blocking-resistant web browsing. It obfuscates the web protocol as a P2P protocol such as VoIP. As CensorSpoofer's source protocol is HTTP, the upstream flow (requested URLs) requires much less bandwidth than the downstream flow (potentially large HTTP responses). Consequently, CensorSpoofer decouples the upstream and downstream flows.

Figure 5.9 shows CensorSpoofer's architecture. A CensorSpoofer client (e.g., a blocked web user) utilizes a low-capacity channel such as email or instant messaging to send HTTP requests to the server (e.g., an HTTP proxy). On the other hand, the server mimics P2P traffic from an oblivious dummy host in order to obfuscate HTTP responses. The server chooses the dummy hosts through random port scanning of Internet IPs.

CensorSpoofer's prototype uses a SIP-based VoIP protocol as the target protocol for obfuscation. A CensorSpoofer client initiates a SIP connection with the CensorSpoofer server by sending a SIP INVITE to the appropriate SIP ID. The CensorSpoofer spoofer replies with a SIP OK message spoofed to look as if its origin is the dummy host. Once the client receives this message, it starts sending encrypted RTP/RTCP packets with random content to the dummy host. At the same time, the spoofer starts sending spoofed, encrypted RTP/RTCP packets to the client ostensibly from the dummy host's address.

To browse a URL, the client sends it through the upstream channel. The spoofer fetches the contents and embeds them in the spoofed RTP packets to the client. To terminate, the client sends a termination signal upstream. The spoofer replies with a spoofed SIP BYE message, and the client sends a SIP OK message and closes the call.

5.3.4 Hybrid Schemes

Hybrid schemes combine two or all three main mechanisms of content-based, pattern-based, and protocol-based obfuscation in order to provide a more resilient protocol type obfuscation. Many of the protocol obfuscation proposals, including some of the example schemes introduced before, are hybrid schemes, while others perform only one obfuscation mechanism. For instance, SkypeMorph [21] (Section 5.2.2.1) performs all content-based, pattern-based, and protocol-based mechanisms, while Obfsproxy [28] (Section 5.3.1.1) deploys only content-based obfuscation. In the following, we introduce StegoTorus [35], which is a hybrid protocol obfuscation scheme.

5.3.4.1 StegoTorus.
Similar to SkypeMorph introduced in Section 5.2.2.1, StegoTorus [35] is a pluggable transport [20] for Tor, which is derived from Obfsproxy [28]. That is, StegoTorus aims at making Tor connections, established between Tor clients and Tor relays, look like other network protocols.

StegoTorus is composed of two main techniques: *chopping* and *steganography*. The chopper performs content-based obfuscation (as introduced in Section 5.3.1); that is, its goal is to foil statistical traffic analysis by modifying packet sizes and timings. The chopper carries Tor traffic over *links* comprised of multiple connections, where each connection is a sequence of blocks, padded and delivered out of order. The second main component of StegoTorus is the steganography module, whose goal is to mimic popular network protocols such as HTTP, Skype, and Ventrilo. That is, the steganography module is responsible for protocol-based obfuscation.

The StegoTorus project [35] implements two variations of the steganography module.

- *Embed steganography:* StegoTorus-Embed aims to mimic a P2P protocol such as Skype or Ventrilo VoIP. It takes as input a database of genuine, previously collected Skype or Ventrilo traffic, and then uses that for shaping the Tor traffic of Tor users. More specifically, embed steganography modifies the timing and sizes of Tor packets based on the timing and size information of the collected network traffic. The Embed module additionally emulates the headers of the target protocols (i.e., content-based obfuscation as introduced in Section 5.3.1), for example, adds required packet headings.

- *HTTP steganography:* The StegoTorus-HTTP module mimics unencrypted HTTP traffic by simulating an HTTP *request generator* and an HTTP *response generator*. These simulators operate by using a pre-recorded trace of HTTP requests and responses collected over actual HTTP sessions. The request generator picks a random HTTP GET request from the trace and hides the payload produced by the chopper in the `<uri>` and `<cookie>` fields of the request by encoding the payload into a modified `base64` alphabet and inserting special characters at random positions to make it look like a legitimate URI or cookie header. The response generator picks a random HTTP response consistent with the request and hides the data in the chapters carried by this response. The StegoTorus implementation uses PDF, SWF, and JavaScript chapters for this purpose.

5.4   COUNTERMEASURES

To foil traffic type obfuscation schemes, various countermeasures [9,37,38] have been proposed that seek one of the following two objectives:

1. Detect the utilization of traffic type obfuscation mechanisms. For instance, state-level censors are interested in detecting the use of censorship resistance systems by their Internet users in order to block users' Internet access.
2. In addition to detecting the use of type obfuscation mechanisms (previous goal), identify the types of obfuscated network protocols. For instance, ISPs are interested in identifying and warning Internet subscribers who use BitTorrent file sharing protocol [3].

Earlier in this chapter (Section 5.3), we classified traffic obfuscation techniques into the three main classes of content-based, pattern-based, and protocol-based schemes. Accordingly, we classify countermeasures to traffic obfuscation techniques into the three categories of *content-based countermeasures*, *pattern-based countermeasures*, and *protocol-based countermeasures* (see Figure 5.10).

5.4.1   Content-Based Countermeasures

Content-based countermeasures aim at detecting traffic obfuscation (and possibly the type of the obfuscated protocol) by looking for content discrepancies in the target traffic. Two examples of content-based countermeasures are provided below.

5.4.1.1   *Missing Skype Headers in Skype Imitators.*   Previously, we introduced SkypeMorph (Section 5.2.2.1) and StegoTorus (Section 5.3.4.1) as two protocol type obfuscation mechanisms that aim at mimicking Skype; that is, they make their network traffic look like Skype. Houmansadr et al. [37] show that these schemes can be identified by using content-based countermeasures. More specifically, Skype uses

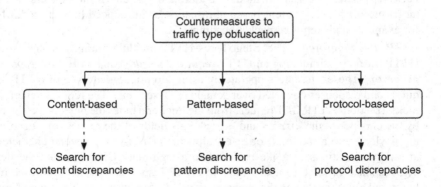

Figure 5.10. Countermeasures to traffic type obfuscation.

special headers, called the start of message (SoM), for its UDP packets [2]. The SoM headers are present in all UDP packets generated by the Skype protocol, and they appear in plaintext, that is, are not encrypted.

Despite Skype being a proprietary protocol, previous work [2,39] has taken steps toward reverse engineering it. In particular, the SoM headers contain two easily recognizable fields of *ID* and *Fun* [2]. *ID*, which uses the first two bytes of the SoM header, uniquely identifies a Skype message. This value is randomly generated by the sender and copied in the receiver's reply. *Fun* is a 5-bit field obfuscated into the third byte of SoM and revealed by applying the `0x8f` bitmask. Previous research [2,39] investigated the values of Fun for different messages. For instance, `0x02`, `0x03`, `0x07`, and `0x0f` indicate signaling messages during the login process and connection management, while `0x0d` indicates a data message, which may contain encoded voice or video blocks, chat messages, or data transfer chunks.

Houmansadr et al. [37] show that both SkypeMorph and StegoTorus obfuscation systems ignore the SoM header in their design. That is, the Skype-like UDP packets generated by SkypeMorph and StegoTorus entirely miss an SoM header, enabling countermeasures that can easily distinguish between imitated and genuine Skype traffic.

5.4.1.2 *Countermeasures Using File-Format Semantics.* As described in Section 5.3.4.1, StegoTorus's "HTTP steganography" module embeds the contents of covert traffic into cover chapters of types PDF, SWF, or JavaScript. StegoTorus does so by using real chapters and replacing specific fields with hidden content. This preserves the file's syntactic validity, but not its semantics. Houmansadr et al. [37] show that these cover chapters can be identified using low-cost content classifiers that use file-format semantics.

In particular, Houmansadr et al. show how to analyze the PDF chapters carrying hidden StegoTorus traffic. StegoTorus uses a fake trace-generator to produce templates for PDF chapters; the generator misses an essential object called *xref table*. In a genuine PDF file, this table follows the `xref` keyword and declares the number and position of various objects. The absence of this table in StegoTorus's imitations is detectable via simple deep-packet inspection at line speed, without any need to reconstruct or parse the file.

Houmansadr et al. additionally demonstrate that adding a fake `xref` table to the PDF file does not fix the detectability issue. This is because there are various software scripts available that are able to check the validity (or invalidity) of a PDF file's `xref` table at low computation cost, with no need of rendering the PDF file. Even if a non-fake `xref` table is generated and used by StegoTorus, there are simple, fast scripts (e.g., the Unix `pdftotext` command) that are able to extract the encoded text from PDF files, which then can be validated through natural language processing techniques.

5.4.2 Pattern-Based Countermeasures

Recent studies [9,32–34] show that many existing pattern-based obfuscation techniques have high overhead, are easily defeatable, or both. In particular, Dyer et al. [9] and Cai et al. [32] evaluate numerous website fingerprinting defenses (including HTTPOS introduced in Section 5.2.1.2 and traffic morphing described in Section 5.3.2.2) and demonstrate the feasibility of countermeasures against them. In the following, we present several example countermeasures.

5.4.2.1 Damerau–Levenshtein Distance.

The Damerau–Levenshtein distance is an information theory tool to measure the distance between two strings, that is, finite sequences of symbols. The Damerau–Levenshtein distance counts the number of operations needed to transform one string to the other, where each operation performs one of the four main tasks of insertion, deletion, substitution of a single character, or transposition of two adjacent characters.

Cai et al. [32] use the Damerau–Levenshtein distance in order to counter several pattern-based obfuscation mechanisms including HTTPOS (Section 5.2.1.2), traffic morphing (Section 5.3.2.2), and padding-based mechanisms (Section 5.2.1.1) such as the randomized pipelining deployed by Tor [40]. They argue that the studied pattern-based obfuscation defenses against traffic analysis are based on ad-hoc heuristics and are likely to fail. Cai et al. additionally use hidden Markov models to extend their webpage classifier into a website classifier.

5.4.2.2 Naïve Bayes Classifier.

Naïve Bayes (NB) is a probabilistic machine learning classifier that applies Bayes' theorem with (naively) strong independence assumptions between the features. Specifically, for a given *feature vector* $Y = (Y_1, Y_2, ..., Y_n)$ of size n, the Bayes theorem returns the conditional probability of $\Pr(\ell_i|Y)$ to be

$$\Pr(\ell_i|Y) = \frac{\Pr(Y|\ell_i)\Pr(\ell_i)}{\Pr(Y)}, \tag{5.5}$$

where ℓ_i is the ith hypothesis out of a set of possible outcomes with size k. The Bayes rule computes $\Pr(\ell_i|Y)$ for each of the $i = \{1, 2, ..., k\}$ potential outcomes and projects the hypothesis with the largest value of $\Pr(\ell_i|Y)$ as the estimated outcome. The naïve form of Bayes rule (NB) assumes that any of the two features Y_s and Y_p of the feature vector Y are independent (for $1 \leq s, p \leq n$ and $s \neq p$), simplifying the Bayes rule to

$$\Pr(\ell_i|Y) = \frac{\Pr(\ell_i)}{\Pr(Y)} \prod_{t=1}^{n} \Pr(Y_t|\ell_i). \tag{5.6}$$

Liberatore and Levine [11] use a naïve Bayes classifier to identify the web destinations browsed over encrypted HTTPS connections. They use the counts of the lengths of the packets sent in each direction of an encrypted HTTPS connection as the feature vector Y. Given that there are 1449 possible packet lengths for an HTTPS packet in each direction, the size of this feature vector is $2 \times 1449 = 2898$. Liberatore and Levine assume equal probability for HTTPS destination, that is, $\Pr(\ell_i) = 1/k$ (k is the number of all possible HTTPS destinations), and use kernel density estimation over the example feature vector to estimate $\Pr(Y|\ell_i)$. The devised algorithm is able to identify the source of encrypted HTTP connections between 60% and 90% of the time, for a bounded set of sources.

Similarly, Herrmann et al. [10] use a *multinominal* naïve Bayes classifier to identify web destinations browsed over encrypted connections. The multinominal naïve Bayes classifier works similar to the naïve Bayes classifier described above, except that it uses the aggregated frequency of the features across all training vectors. That is, the features are evaluated using a normalized counting of the packet sizes as opposed to a raw counting.

5.4.2.3 Support Vector Machine Classifier.
A support vector machine (SVM) [41] is a supervised machine learning algorithm that can be used as a classifier for observed data. Given a set of training data points each tagged as belonging to one of the two possible categories, SVM represents each of the examples as points in a high-dimensional space and derives a hyperplane that maximally separates the points belonging to the two categories.

Panchenko et al. [42] suggest the use of SVM classifiers to identify websites browsed over encrypted anonymity systems such as Tor [4]. Their proposed classifier utilizes a wide range of features of network flows pertaining to the volume, time, and direction of the traffic. This includes fine-grained features such as counts of packet lengths, and coarse-grained features such as the total number of transmitted packets. All of the used features are rounded and all TCP acknowledgment packets are removed to minimize noise during the training. The proposed scheme achieves a surprisingly high true detection rate of around 73% for a very small false negative rate.

5.4.2.4 Standard Deviation.
FreeWave, introduced in Section 5.2.2.2, obfuscates the underlying network protocol by encoding cover messages into audio signals and sending them over encrypted VoIP connections. Geddes et al. [38] show that FreeWave connections build over Skype can be distinguished from genuine Skype calls by performing statistical traffic analysis. Such countermeasures stem from the fact that Skype's audio codec, SILK [43], is a variable-bit length codec, which makes the sizes and timings of Skype packets depend on the spoken conversation. Geddes et al. particularly show that the standard deviation of packet lengths can distinguish a genuine Skype connection from one carrying FreeWave's data-modulated audio.

Note that such attacks were acknowledged in the original FreeWave paper [22], where they proposed three countermeasures.

1. To deploy FreeWave on a VoIP protocol that, unlike Skype, uses a fixed-bit rate codec (e.g., the widely used G.7xx[4] codec series). This will make traffic patterns independent of the underlying conversation.
2. To superimpose the modulated audio of FreeWave with pre-recorded human conversations.
3. To tunnel cover traffic in the video content of a VoIP conversation as a video has less dependence on the underlying content compared with audio.

---

[4] http://www.voip-info.org/wiki/view/Codecs.

TABLE 5.1. Passive protocol-based countermeasures to detect imitators of the Skype protocol.

Countermeasure	SkypeMorph	StegoTorus
Skype HTTP update traffic	Not defeated	Defeated
Skype login traffic	Not defeated	Defeated
Periodic message exchanges	Defeated	Defeated
Typical Skype client behavior	Defeated	Defeated
TCP control channel	Defeated	Defeated

Reproduced from [24] with permission of IEEE.

5.4.3 Protocol-Based Countermeasures

Protocol-based countermeasures [37,38] look for abnormalities in the traffic imitated by a protocol-based obfuscation scheme. They include *passive* countermeasures, which passively monitor network traffic for discrepancies, and *active* countermeasures, which perturb suspected traffic looking for revealing behavior.

5.4.3.1 Passive Countermeasure Examples. SkypeMorph (Section 5.2.2.1) and StegoTorus (Section 5.3.4.1) obfuscate their traffic by imitating Skype's protocol. Despite the Skype protocol being proprietary, it has been reverse engineered in previous work [2,39,44], resulting in various Skype identification tests [2,39] that are used by ISPs and enterprise networks to detect (and sometimes block) the use of Skype. Houmansadr et al. [37] show that such tests can be used as countermeasures to identify protocol-based obfuscations schemes that—incompletely—imitate Skype. Table 5.1 shows some of the tests that can be used to identify SkypeMorph and StegoTorus.

Houmansadr et al. [37] argue that one can combine the tests listed in Table 5.1 into a hierarchical detection tool to build more advanced countermeasures. In fact, similar tools have been proposed for real-time detection of Skype traffic [45,46], including line-rate detectors by Pláček [39], who used these tests in an NfSen[5] plug-in, and by Adami et al. [8].

5.4.3.2 Active Countermeasure Examples. In the following, we describe several active countermeasures proposed by Houmansadr et al. [37] that target protocol-based obfuscation systems that mimic Skype.

MANIPULATING SKYPE CALLS. This countermeasure tampers with a Skype connection by dropping, reordering, and delaying packets or modifying their contents, and then observes the endpoints' reaction. These changes are fairly mild and can occur naturally; thus, they do not drastically affect genuine Skype connections.

When UDP packets are dropped in a genuine Skype call, there is an immediate, very noticeable increase in the activity on the TCP control channel that accompanies the

---

[5] http://nfsen.sourceforge.net/.

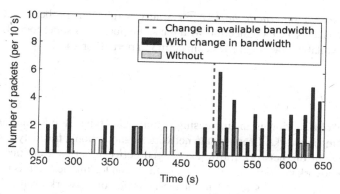

Figure 5.11. Skype TCP activity with and without changes in bandwidth. (Reproduced from [37] with permission of IEEE.)

main UDP connection (see Figure 5.11). Houmansadr et al. conjecture that this is caused by Skype endpoints renegotiating connection parameters due to perceived changes in network conditions.

Houmansadr et al. argue that it is *extremely* difficult for an imitator of Skype to convincingly imitate dynamic dependencies between network conditions and Skype's control traffic; hence, this can be used as a countermeasure to detect such protocol-based obfuscation mechanisms. This active countermeasure does not adversely affect normal Skype users. Dropping a few packets does not disconnect the call, but only degrades its quality for a short period of time. Geddes et al. [38] propose similar countermeasures that aim at disrupting (as opposed to identifying) Skype's protocol-based obfuscation systems by strategically dropping Skype packets.

MANIPULATING THE TCP CONTROL CHANNEL OF SKYPE. The countermeasures introduced above assert that perturbing Skype's main UDP connection causes observable changes in the TCP control channel. Alternatively, Houmansadr et al. [37] show that perturbing the TCP channel of Skype can cause observable changes in the UDP connection, which could be used to identify imitations of Skype.

1. *Closing the TCP connection:* Closing the TCP channel (e.g., by sending an RST packet) causes genuine Skype nodes to immediately end the call. A typical imitator of Skype is likely not to mimic this behavior because their fake TCP channel may have no relationship to the actual call.

2. *Withholding or dropping selected TCP packets:* The TCP connection of Skype sends a packet every 30–60 s, or when network conditions change. Tampering with these packets causes observable changes in the genuine UDP channel, but not on an imitated one.

3. *Triggering a supernode probe:* A Skype client keeps a TCP connection with its supernode [2]. If this connection is closed, a genuine client immediately launches a UDP probe [44] to search for new supernodes.

4. *Blocking a supernode port:* After a successful UDP probe, a genuine client establishes a TCP connection with the same port of its supernode. If this port is not available, the client tries connecting to port 80 or 443 [44].

5.5 SUMMARY

Traffic type obfuscation is an increasingly important type of information hiding in communication networks. In contrast to network steganography, traffic type obfuscation hides the *type* of the network traffic exchanged between network entities, that is, the underlying network protocols. Traffic type obfuscation techniques do this by modifying contents of traffic, patterns of traffic, or the behavior of network protocols. They are utilized mainly for blocking resistance and privacy protection.

REFERENCES

1. Y.-D. Lin, C.-N. Lu, Y.-C. Lai, W.-H. Peng, and P.-C. Lin. Application classification using packet size distribution and port association. *Journal of Network and Computer Applications*, 32(5):1023–1030, 2009.
2. D. Bonfiglio, M. Mellia, M. Meo, D. Rossi, and P. Tofanelli. Revealing Skype traffic: when randomness plays with you. *ACM SIGCOMM Computer Communication Review*, 37(4):37–48, 2007.
3. K. Bauer, D. McCoy, D. Grunwald, and D. Sicker. BitStalker: accurately and efficiently monitoring BitTorrent traffic. In *1st IEEE International Workshop on Information Forensics and Security (WIFS 2009)*, pp. 181–185. IEEE, 2009.
4. R. Dingledine, N. Mathewson, and P. Syverson. Tor: the second-generation onion router. In *USENIX Security*, 2004.
5. P. Winter and S. Lindskog. How the Great Firewall of China is blocking Tor. In *FOCI*, 2012.
6. T. Wilde. Knock knock knockin' on bridges' doors. https://blog.torproject.org/blog/knock-knock-knockin-bridges-doors, 2012.
7. Ten ways to discover Tor bridges. https://blog.torproject.org/blog/research-problems-ten-ways-discover-tor-bridges.
8. D. Adami, C. Callegari, S. Giordano, M. Pagano, and T. Pepe. Skype-Hunter: a real-time system for the detection and classification of Skype traffic. *International Journal of Communication Systems*, 25(3):386–403, 2012.
9. K. P. Dyer, S. E. Coull, T. Ristenpart, and T. Shrimpton. Peek-a-boo, I still see you: why efficient traffic analysis countermeasures fail. In *2012 IEEE Symposium on Security and Privacy (SP)*, pp. 332–346. IEEE, 2012.
10. D. Herrmann, R. Wendolsky, and H. Federrath. Website fingerprinting: attacking popular privacy enhancing technologies with the multinomial naïve-Bayes classifier. In *Proceedings of the 2009 ACM Workshop on Cloud Computing Security*, pp. 31–42. ACM, 2009.
11. M. Liberatore and B. N. Levine. Inferring the source of encrypted HTTP connections. In *Proceedings of the 13th ACM Conference on Computer and Communications Security*, pp. 255–263. ACM, 2006.

12. G. D. Bissias, M. Liberatore, D. Jensen, and B. N. Levine. Privacy vulnerabilities in encrypted HTTP streams. In *Privacy Enhancing Technologies*, pp. 1–11. Springer, 2006.

13. A. Hintz. Fingerprinting websites using traffic analysis. In *Privacy Enhancing Technologies*, pp. 171–178. Springer, 2003.

14. Q. Sun, D. R. Simon, Y.-M. Wang, W. Russell, V. N. Padmanabhan, and L. Qiu. Statistical identification of encrypted web browsing traffic. In *Proceedings of the 2002 IEEE Symposium on Security and Privacy*, pp. 19–30. IEEE, 2002.

15. S. Chen, R. Wang, X. Wang, and K. Zhang. Side-channel leaks in web applications: a reality today, a challenge tomorrow. In *2010 IEEE Symposium on Security and Privacy (SP)*, pp. 191–206. IEEE, 2010.

16. C. V. Wright, S. E. Coull, and F. Monrose. Traffic morphing: an efficient defense against statistical traffic analysis. In *Network and Distributed System Security Symposium (NDSS)*, 2009.

17. X. Luo, P. Zhou, E. W. Chan, W. Lee, R. K. Chang, and R. Perdisci. HTTPOS: sealing information leaks with browser-side obfuscation of encrypted flows. In *Network and Distributed System Security Symposium (NDSS)*, 2011.

18. M. Dusi, M. Crotti, F. Gringoli, and L. Salgarelli. Tunnel Hunter: detecting application-layer tunnels with statistical fingerprinting. *Computer Networks*, 53(1):81–97, 2009.

19. How governments have tried to block Tor. https://svn.torproject.org/svn/projects/presentations/slides-28c3.pdf.

20. Tor: pluggable transports. https://www.torproject.org/docs/pluggable-transports.html.en.

21. H. Moghaddam, B. Li, M. Derakhshani, and I. Goldberg. SkypeMorph: protocol obfuscation for Tor bridges. In *Proceedings of the ACM Conference on Computer and Communications Security*, 2012.

22. A. Houmansadr, T. Riedl, N. Borisov, and A. Singer. I want my voice to be heard: IP over Voice-over-IP for unobservable censorship circumvention. In *Network and Distributed System Security Symposium (NDSS)*, 2013.

23. R. Fielding, J. Gettys, J. Mogul, H. Frystyk, L. Masinter, P. Leach, and T. Berners-Lee. Hypertext Transfer Protocol—HTTP/1.1. RFC 2616, June 1999.

24. M. Leech, M. Ganis, Y. Lee, R. Kuris, D. Koblas, and L. Jones. SOCKS Protocol Version 5. RFC 1928, April 1996.

25. V. Paxson. Bro: a system for detecting network intruders in real-time. *Computer Networks*, 31(23):2435–2463, 1999.

26. C. Inacio and B. Trammell. Yaf: yet another flowmeter. In *Proceedings of the Large Installation System Administration Conference (LISA)*, 2010.

27. L. Deri. nprobe: an open source netflow probe for gigabit networks. In *Proceedings of TERENA Networking Conference*, 2003.

28. A simple obfuscating proxy. https://www.torproject.org/projects/obfsproxy.html.en.

29. B. Wiley. Dust: a blocking-resistant internet transport protocol. Technical Report, School of Information, University of Texas at Austin, 2011.

30. K. Dyer, S. Coull, T. Ristenpart, and T. Shrimpton. Protocol misidentification made easy with format-transforming encryption. In *Proceedings of the ACM Conference on Computer and Communications Security*, 2013.

31. M. Bellare, T. Ristenpart, P. Rogaway, and T. Stegers. Format-preserving encryption. In *Selected Areas in Cryptography*, pp. 295–312. Springer, 2009.

32. X. Cai, X. C. Zhang, B. Joshi, and R. Johnson. Touching from a distance: website finger-printing attacks and defenses. In *Proceedings of the 2012 ACM Conference on Computer and Communications Security*, pp. 605–616. ACM, 2012.

33. X. Cai, R. Nithyanand, T. Wang, R. Johnson, and I. Goldberg. A systematic approach to developing and evaluating website fingerprinting defenses. In *Proceedings of the ACM Conference on Computer and Communications Security*, 2014.

34. M. Juarez, S. Afroz, G. Acar, C. Diaz, and R. Greenstadt. A critical evaluation of web-site fingerprinting attacks. In *Proceedings of the 21st ACM Conference on Computer and Communications Security (CCS 2014)*, 2014.

35. Z. Weinberg, J. Wang, V. Yegneswaran, L. Briesemeister, S. Cheung, F. Wang, and D. Boneh. StegoTorus: a camouflage proxy for the Tor anonymity system. In *Proceedings of the ACM Conference on Computer and Communications Security*, 2012.

36. Q. Wang, X. Gong, G. Nguyen, A. Houmansadr, and N. Borisov. CensorSpoofer: asymmetric communication using IP spoofing for censorship-resistant web browsing. In *Proceedings of the ACM Conference on Computer and Communications Security*, 2012.

37. A. Houmansadr, C. Brubaker, and V. Shmatikov. The parrot is dead: observing unobservable network communications. In *34th IEEE Symposium on Security and Privacy*, Oakland, CA, 2013.

38. J. Geddes, M. Schuchard, and N. Hopper. Cover your ACKs: pitfalls of covert channel censorship circumvention. In *Proceedings of the 2013 ACM SIGSAC Conference on Computer and Communications Security (CCS)*, pp. 361–372. ACM, 2013.

39. L. Ptáček. Analysis and detection of Skype network traffic. Master's thesis, Masaryk University, 2011.

40. M. Perry. Experimental defense for website traffic fingerprinting. https://blog.torproject.org/blog/experimental-defense- website-traffic-fingerprinting, September 2011.

41. C. Cortes and V. Vapnik. Support-vector networks. *Machine Learning*, 20(3):273–297, 1995.

42. A. Panchenko, L. Niessen, A. Zinnen, and T. Engel. Website fingerprinting in onion routing based anonymization networks. In *Proceedings of the 10th Annual ACM Workshop on Privacy in the Electronic Society*, pp. 103–114. ACM, 2011.

43. S. Jensen, K. Vos, and K. Soerensen. SILK speech codec. Technical Report draft-vos-silk-02.txt, IETF Secretariat, Fremont, CA, September 2010.

44. S. Baset and H. Schulzrinne. An analysis of the Skype peer-to-peer Internet telephony protocol. In *INFOCOM*, 2006.

45. D. Bonfiglio and M. Mellia. Tracking down Skype traffic. In *INFOCOM*, 2008.

46. E. Freire, A. Ziviani, and R. Salles. On metrics to distinguish Skype flows from HTTP traffic. *Journal of Network and Systems Management*, 17(1–2):53–72, 2009.

6

NETWORK FLOW WATERMARKING

Absence of understanding does not warrant absence of existence.

—Avicenna (Ibn Sina)

Network steganography, introduced in Chapter 3, is the art of hiding information into the communication process. Flow watermarking is a particular type of network steganography, that is, it hides information into network traffic. Flow watermarking is distinct from other types of network steganography in two ways: First, the information hidden by a flow watermark has no value by itself, and is inserted as a metadata to augment its carrying network flow. For instance, a hidden watermark may convey some information about the network origin of its carrying network flow. On the other hand, the information hidden by regular network steganography is valuable by itself, and is independent of its carrying network flow.

Second, flow watermarks hide information only into the *communication patterns* of traffic, whereas other types of network steganography may additionally make use of *communication content* for hiding information. The communication patterns commonly used for flow watermarking include traffic characteristics such as packet timings, packet sizes, packet counts, and flow rates.

Information Hiding in Communication Networks: Fundamentals, Mechanisms, Applications, and Countermeasures,
First Edition. Wojciech Mazurczyk, Steffen Wendzel, Sebastian Zander, Amir Houmansadr, and Krzysztof Szczypiorski.
© 2016 by The Institute of Electrical and Electronics Engineers, Inc. Published 2016 by John Wiley & Sons, Inc.

Flow watermarking is primarily used for *linking network flows* in application scenarios where packet contents are striped of all linking information. The flows of interest in these applications are commonly relayed by several low-latency nodes, where each node adds a layer of encryption to the traffic payload making network flows unlinkable through correlating their packet headers and payloads. In this case, flow watermarking is a powerful tool for linking network flows as it correlates the communication patterns of network flows that are not significantly changed due to encryption and relaying, as opposed to correlating their contents. An example application scenario for flow watermarking is in intrusion detection [1–3]. In this application, flow watermarking is used to link network flows belonging to the same attack session, for example, different segments of a TCP connection relayed by a public network proxy. This will enable the traffic analyzers to trace back to the originator of an attack, disguising her identity behind the public network proxy. Another example scenario for flow watermarking is in compromising anonymity systems [4–6]. In this scenario, privacy invaders use flow watermarking to link communicating parities in the anonymity network by linking their anonymized traffic flows.

Network flow watermarking is preceded by a passive form of network traffic analysis, called passive flow correlation [7–11]. In passive flow correlation, network flows are linked merely by observing their traffic patterns, as opposed to flow watermarking in which flow patterns are manipulated before the correlation. The main shortcoming of passive traffic analysis is not being *scalable* to large-size applications. Additionally, passive correlation schemes usually suffer from high rates of false-positive error rates due to the intrinsic correlation between (unrelated) network flows. Flow watermarking aims at addressing these two shortcomings of passive traffic analysis schemes.

6.1 PRINCIPLES, DEFINITIONS, AND PROPERTIES

We start by reviewing network traffic analysis and its utilization in linking network flows, that is, through flow watermarks. This is then followed by an overview of the main building blocks of flow watermarking systems, and their main features.

6.1.1 Network Traffic Analysis

Traffic analysis is inferring *sensitive information* from *communication patterns*, instead of traffic content. The communication patterns used by traffic analysis include flow characteristics such as packet timings, flow rates, packet sizes, etc. Depending on the particular networking application, such patterns are usually not drastically altered due to the use of encryption or the noise from the network. As a result, traffic analysis is significantly useful in networking applications where encryption prevents the analysis of packet contents. With the increasing use of encryption and other evasion techniques on network flows, which strip packet contents of any information that can be used for traffic analysis, traffic analysis is becoming more relevant and applicable to various security and privacy problems.

Applications of Traffic Analysis. As mentioned above, traffic analysis is a tool for inferring ''sensitive information'' by analyzing communication patterns. Depending on the particular type of information inferred by traffic analysis, it can be used in various applications. As one example, traffic analysis can be used to derive sensitive information from voice-over-IP (VoIP) conversations such as Skype calls. In particular, researchers have designed traffic analysis tools that are able to determine the language spoken in a VoIP conversation [12], disclose the identity of the callees [13,14], or even, more importantly, unmask the contents of VoIP conversations [15,16].

The traffic analysis methods discussed in this chapter target linkability of network flows. That is, the ''sensitive information'' inferred by this class of traffic analysis techniques is the ''relation'' between network flows. Such traffic analysis tools aim at linking network flows that are otherwise unlinkable due to the use of encryption or other evasion techniques.

6.1.2 Linking Network Flows

Linking network flows is an important, challenging problem in various networking applications. For instance, linking network flows can help forensics analysis to identify cybercriminals who relay their network traffic through network proxies, known as stepping stones [7,8,17], to evade detection. Consider a cybercriminal who relays his traffic through a previously compromised machine to escape detection, as shown in Figure 6.1. Such a cybercriminal can be detected if some network entities on the path (for example, the border router and the stepping stone detector router in Figure 6.1) have a reliable tool to link different parts of the attack traffic, that is, flows 2 and 5 in the figure.

Traffic analysis is a powerful tool for linking network flows, especially in the presence of encryption and other evasion techniques such as traffic proxying.

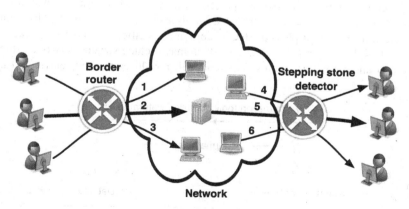

Figure 6.1. Linking network flows for the detection of stepping stone attacks. Flows numbered 2 and 5 are part of a stepping stone attack, while the other flows are benign.

Traffic analysis methods used for linking network flows are broadly classified into two categories, described in the following.

6.1.2.1 Passive Linking (Flow Correlation). Traditionally, linking network flows is done in a passive manner, known as *flow correlation*. Passive flow correlation schemes link network flows by *merely* monitoring the network traffic, trying to detect related network flows by correlating features such as packet timings, sizes, and counts [7–11]. Flow correlation schemes have been proposed to link network flows in various applications, including the detection of stepping stone attacks and compromising anonymous communications.

6.1.2.2 Active Linking (Flow Watermarking). More recently, an active approach called *flow watermarking* has been studied for traffic analysis. Network flow watermarking is the art of *hiding* information within the communication patterns of network flows, for example, packet timing characteristics. The hidden information can be used to link network flows, to identify the source of a network connection, or to exchange covert messages.

In this approach, traffic patterns of one flow (usually packet timings) are *actively* modified to contain a special pattern. If the same pattern is later found on another flow, the two are considered linked. Watermarking significantly reduces the computation and communication costs of traffic analysis, and also leads to more precise detection with fewer false positives. Watermarking has been considered for different applications of traffic analysis, for example, detection of stepping stones [1–3] and compromising anonymous networks [4–6]. More recently, watermarking has also been studied as a traceback mechanism for botnets [18,19].

6.1.3 Building Blocks of Flow Watermarking Systems

Watermarks provide a content-independent way to tag traffic so that correlated flows can later be recognized. Figure 6.2 shows the general model of a network flow watermarking scheme. A network flow passing through a *watermarker* gets *watermarked* by changing the timing information of packets, that is, applying specific delays on the packets. The flow then travels along a noisy channel, which may include various networks, stepping stones, and anonymizing systems. This channel introduces further delays, and might

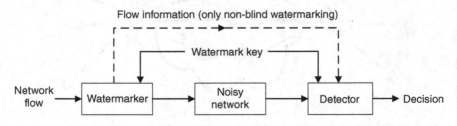

Figure 6.2. General model of network flow watermarking.

also drop, reorder, or duplicate packets or repacketize the flow. After the channel, the flow arrives at the watermark *detector*, which inspects it for the inserted watermark pattern. The pattern encoding is based upon a secret watermark key, shared between the watermarker and the detector. In a *blind* scheme, this is the only information the watermarker and detector exchange; in a *non-blind* scheme, the watermarker also sends information about the watermarked flow to the detector through an out-of-band channel.

Watermarking has two main advantages over passive flow correlation schemes.

1. *Superior linking performance:* Flow watermarking can provide lower false-positive rates than passive flow correlation analysis in linking network flows. This is because passive analysis works by correlating flow patterns that can be intrinsically correlated among nonlinked flows, whereas active traffic analysis correlates artificial patterns that are tailored to be highly uncorrelated among nonlinked flows. For instance, passive traffic analysis of FTP traffic is likely to result in high rates of false positives due to the intrinsic correlation between (even nonrelated) FTP flows.

2. *Better scalability:* The main shortcoming of passive traffic analysis is not being *scalable* to large-size applications. For a network with n ingress flows and m egress flows, passive traffic analysis needs to perform $O(mn)$ flow correlations, and $O(n)$ flows need to be communicated among traffic analysis entities. Flow watermarking reduces[1] the computational overhead to $O(n)$ (compared to $O(mn)$ for passive analysis) and the communication overhead to $O(1)$ (compared to $O(n)$ in the case of passive analysis), since the information used for flow linking, that is, the watermark, is self-contained in each flow. In other words, flow watermarking provides *lower computation overhead*, as well as *lower communication overhead* compared to passive flow correlation.

6.1.4 Features of Flow Watermarks

Watermarking schemes are evaluated based on several features.

- *Detection accuracy:* A detector who possesses the watermark key should be able to accurately detect watermarks. Watermark detection accuracy is measured by the two metrics of *miss rate* and *false-positive rate*. The miss rate is the rate of watermark flows not successfully detected by the detector. On the other hand, the false-positive rate is the rate of nonwatermarked flows mistakenly declared as watermarked by the detectors. Detection accuracy is inversely proportional to both miss rate and false-positive rate.

- *Robustness:* A watermark should be able to survive perturbations to network traffic, such as delays/jitter, packet drops, retransmissions, and repacketization.

---

[1] This does not apply to non-blind flow watermarking, as will be explained later.

In other words, watermark detection should be accurate (as defined above) in the presence of natural communication noise.

- *Active robustness:* A watermark should be able to survive intentional modifications to network traffic introduced by an attacker,[2] under some bounded model (a common one is a total delay bound). For some application scenarios of watermarks, an active attack is either impractical or inefficient; therefore, watermarks designed for such purposes do not consider active robustness. For instance, the watermarks studied for compromising anonymity systems do not aim resistance to active perturbations. This is because watermark attackers in this scenario are anonymity proxies (e.g., Tor routers) who refrain from applying extra perturbation on anonymous communication traffic.

- *Invisibility:* A watermark must introduce pattern modifications (e.g., timing delays) that do not significantly interfere with the flow's performance; this is particularly important since most watermarked flows are benign flows. That is, watermark insertion should be imperceptible to legitimate users. Additionally, the presence of a watermark should not be visible to an attacker who does not have the secret watermark key, even if the attacker specifically looks for it.

6.2 APPLICATIONS OF FLOW WATERMARKS

Flow watermarking is a powerful tool that can be used to link network flows in various networking applications. Specifically, network flow watermarking can be used in any networking application that satisfies the following two conditions:[3]

1. *Linking is not feasible using traffic content:* Trivially, one can link network flows by comparing packet contents. The use of traffic content (e.g., IP packet headers and TCP payloads) for linking flows is much more efficient than doing so using traffic patterns (i.e., flow watermarking), as it generates much smaller rates of false-positive and false-negative errors. So, flow watermarking should only be used when traffic content is stripped of any information that can be used for linking flows, for example, due to the use of encryption.

2. *Communication channel incurs low perturbation:* The basis of flow watermarking is to embed an invisible tag into a network flow by artificially perturbing its pattern. For flow watermarking to be successful, such artificial perturbation, that is, the watermark, should remain on the network flow while it traverses the communication environment, for example, a computer network. That is, the watermark should resist the perturbations from the communication environment, for example, the timing perturbations applied to the packets of a watermarked

---

[2] An attacker is a network end point who aims at removing flow watermarks from network traffic.

[3] Note that passive flow correlation can also be used in such applications, resulting in a worse performance (e.g., accuracy) and less scalability, as discussed earlier.

flow. As an example, timing-based flow watermarks have been proposed for low-latency anonymity systems like Tor [20], but not for high-latency anonymity systems like Mixminion [21].

In the following, we briefly describe different application scenarios for network flow watermarking. We start by describing the use of flow watermarking for the two applications of stepping stone detection and compromising anonymity networks. The main body of research on flow watermarking encompasses these two applications. This is followed by two other, more recently proposed, application scenarios for network flow watermarks.

6.2.1 Detection of Stepping Stone Attacks

A stepping stone is a host that is used to relay traffic through a network to another remote destination. Stepping stones are used to disguise the true origin of an attack. Detecting stepping stones can help trace attacks back to their true source. Also, stepping stones are often indicative of a compromised machine. Thus, detecting stepping stones is a useful part of enterprise security monitoring.

Generally, stepping stones are detected by noticing that an outgoing flow from an enterprise matches an incoming flow. For example, in Figure 6.1, flow 2 will have the same characteristics as flow 5, allowing someone to link them by correlating such characteristics. Since the relayed connections are often encrypted (using SSH [22], for example), only characteristics such as packet sizes, counts, and timings are available for such detection. And even these are not perfectly replicated from an incoming flow to an outgoing flow, as they are changed by padding schemes, retransmissions, and jitter. As a result, traffic analysis is used to detect correlations among the incoming and outgoing flows, that is, to link them. Flow watermarks have been studied [1–3] as an effective tool to detect stepping stone attacks. As Figure 6.3 shows, a watermarker can manipulate traffic characteristics of network flows entering the network, which are then looked for by some watermark detector.

6.2.2 Compromising Anonymous Communication

At a very high level, an anonymous system maps a number of input flows to a number of output flows while hiding the relationship between them, as shown in Figure 6.4. The internal operation can be implemented in various manners, for example, using onion routing [23] or a simple proxy [24]. The goal of an attacker, then, is to link an incoming flow to an outgoing flow (or vice versa).

A watermark can be used to defeat anonymity protection by marking certain input flows and watching for marks on the output flows (see Figure 6.4). For example, a malicious website might insert a watermark in all flows from the site to the anonymizing system. A cooperating attacker who can eavesdrop on the link between a user and the anonymous system can then determine if the user is browsing the site or not. Similarly, a compromised entry router in Tor can watermark all of its flows, and cooperating exit routers or websites can detect this watermark.

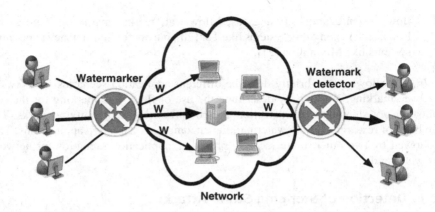

Figure 6.3. Using flow watermarks to detect stepping stone attacks.

Note that this does not enable a fundamentally new attack on low-latency anonymous systems: It has been long known [25] that if an attacker can observe a flow at two points, he can determine if the flow is the same, unless cover traffic is used. (In fact, deployed low-latency systems such as Onion Routing [23], Freedom [26], and Tor [20] have all opted to forego cover traffic due to its being expensive, hoping instead that it will be difficult for an attacker to observe a significant fraction of incoming and outgoing flows.) However, watermarking makes the attack much more efficient, as demonstrated in previous research [4–6].

6.2.3 Detection of Centralized Botnets

A *botnet* is a network of compromised machines, *bots*, that is controlled by one or more *botmasters* to perform coordinated malicious activity. Botnets are among the most serious threats in cyberspace due to their large size [27]. This enables the bots to carry out various attacks, such as distributed denial of service, spam, and identity theft, on a massive scale.

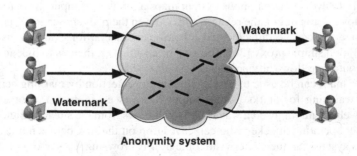

Figure 6.4. A system for anonymous communications.

Botnets are controlled by means of a command-and-control (C&C) channel. A common approach is to use an Internet Relay Chat (IRC) channel for C&C: All the bots and a botmaster join a channel and the botmaster uses the channel to broadcast commands, with responses being sent back via broadcast or private messages to the botmaster. The IRC protocol is designed to support large groups of users and a network of servers to provide scalability and resilience to failures; it thus forms a good fit for providing a C&C infrastructure. Because of their simple design and deployment, IRC botnets have been widely used by cybercriminals since 2001 [28]. Some botnets use a more advanced structure, with bots communicating directly with each other in a peer-to-peer fashion, but recent studies show that many existing botnets use the IRC model because of its simple-yet-effective structure [28,29].

The use of encryption by IRC servers foils linking flows using packet contents. Additionally, botmasters typically use stepping stones [8] to hide their true location (see Figure 6.5). To tackle this, researchers [19,30] have proposed the use of flow watermarks to detect botmasters and infected bot machines in centralized botnets, for example, IRC-based botnets. In this approach, an organization runs an infrastructure that captures bots across the Internet, for example, using a Honeynet [31]. The captured bots remain in the control of their botmasters, that is, they continue their communication over the C&C channel. The organization hosting such captured bots will insert a particular type of flow watermarks into the communication between the captured bots and the IRC server being used for the C&C channel (see Figure 6.5).

The watermark will be visible on all connections between the bots and the IRC server. It will likewise appear in the connection from the botmaster to the IRC server. The watermark can be used to detect such stepping stones and aid in botmaster traceback. As illustrated in Figure 6.5, this enables a network administrator who has obtained the watermarking key from the watermarking organization to identify bot-infected machines in her network territory (e.g., an ISP), as well as the botmasters. The inserted watermark

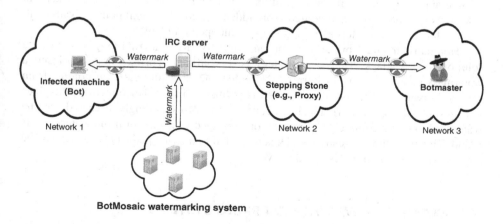

Figure 6.5. A botnet traceback system [30] using flow watermarks.

patterns can be recognized simply by observing the timings of the packets in a given flow; thus the detection can be carried out at a large scale by border routers.

6.2.4 Mitigating Loopback Attacks in Tor

Section 6.2.2 described the use of flow watermarking as a privacy-invasive tool against an anonymity systems such as Tor [20]. This section presents a privacy-enhancing application of watermarks, which aims at foiling a specific type of attack against Tor.

Evans et al. [32] demonstrated an attack on Tor that uses active probing to detect which Tor routers are used to forward a particular tunnel, thus breaking anonymity. Unlike watermarks or passive traffic analysis, their attack works even when the routers being used are *not* under the control or observation by the adversary. The basis of the attack comes from an earlier congestion attack, explored by Murdoch and Danezis [33]. However, a key feature of the new attack is the use of bandwidth amplification to create sufficient congestion so as to make this attack practical on today's Tor network.

The bandwidth amplification exploits the fact that paths in Tor can be constructed to have an arbitrary length. This, coupled with the fact that each hop on a path knows only the previous and the next hops, makes it easy to construct a path that loops through a set of routers many times. This, in turn, ensures that a single packet sent by a user will result in k packet transmissions at each of the routers in the loop, for near-arbitrary values of k.

A potential defense described by Evans et al. is to modify the Tor protocol to restrict the number of circuit extensions it allows, and thus the maximum path length. However, they point out that this is not sufficient to completely prevent such congestion attacks, as loops can still be created by going outside the Tor network and then returning. In particular, a malicious client can create a Tor circuit that connects to a non-Tor proxy, which then connects *back* to Tor pretending to be a new client, and so on and so forth. Iterating this process yields the same functionality as the long-path attack. Although a naive approach may be foiled by exit and entrance policies in Tor, the attacker can instead use proxies, other anonymizers, or hidden Tor entry and exit points. Evans et al. leave defense to such external routing loops as an open problem.

Houmansadr and Borisov [34] propose the SWIRL watermarking system as a solution to this attack. The basic strategy is to configure Tor exit nodes to insert a watermark on all outgoing TCP traffic. Note that this labels the traffic as coming from Tor, but given that the list of exit nodes is published in the Tor directory, this does not significantly degrade privacy. Each entry guard, correspondingly, tries to detect the watermark on an incoming TCP connection and rejects the stream if the watermark is found. This way, the congestion attack is restricted to internal paths only, which can be solved using the solution described above.

6.3 EXAMPLE FLOW WATERMARKING SYSTEMS

In this section, we introduce several example systems for flow watermarking.

6.3.1 Types of Flow Watermarks

Figure 6.2 shows the general model of flow watermarking schemes. Flow watermarking schemes can be classified based on the following two criteria.

- *The information shared between watermarker and detector:* We call a flow watermarking scheme to be a *blind scheme* if the only information communicated between its watermarkers and detectors is the secret watermark key (but no information about the intercepted flows). On the other hand, we call a flow watermarking scheme to be a *non-blind scheme* if the watermarking entities exchange some information about the network flows being watermarked, in addition to the secret watermark key. We will provide examples of each of the classes in the rest of this chapter.

- *The features used for watermarking:* Flow watermarks use a variety of flow patterns for their operation. While the majority of the proposed flow watermarks use the *packet timings* pattern for watermarking [1,3–5,34], there are proposals that utilize other patterns such as *packet sizes* [35] and *flow rates* [6]. Two common types of timing-based watermarks are *interpacket delay (IPD)-based* schemes [1,3] and *interval-based* schemes [4,5,34]. Interval-based flow watermarks apply their modifications in timing intervals of a target flow, whereas in IPD-based schemes, watermark modifications are applied to pairs of consecutive packets. In the following, we will introduce example schemes for each of the classes.

6.3.2 Interval-Centroid-Based Watermarking (ICBW)

This section introduces an interval-based timing watermark, called the interval-centroid-based watermark (ICBW) [5]. The scheme is based on dividing the stream into intervals of equal lengths, using two parameters: o, the offset of the first interval, and T, the length of each interval. A subset of $2n$ of these intervals is randomly selected that is subsequently randomly divided into two further subsets A and B each consisting of $n = rl$ intervals. Each of the sets A and B are randomly divided into l subsets denoted by $\{A_i\}_{i=1}^{l}$ and $\{B_i\}_{i=1}^{l}$ each consisting of r intervals. The ith watermark bit is encoded using the sets $\{A_i, B_i\}$. Therefore, a watermark of length l can be embedded in the flow. Figure 6.6 depicts the random selection and grouping of time intervals of a packet flow for watermark insertion.

The watermarker and detector agree on the parameters o and T, and use a random number generator (RNG) and a seed s to randomly select and assign intervals for watermark insertion. To keep the watermark transparent, all of these parameters are kept secret. Depending on whether the ith watermark bit is 1 or 0, the watermarker delays the arrival times of the packets at the interval positions in sets A_i or B_i, respectively, by a maximum of a. Figure 6.7 illustrates the effect of this delaying strategy over the distribution of packets arrival time in an interval of size T (this operation is called "squeezing" by Wang *et al.*)

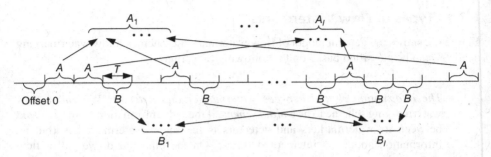

Figure 6.6. Random selection and assignment of time intervals of a packet flow for watermark insertion.

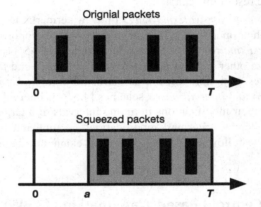

Figure 6.7. Distribution of packets arrival time in an interval of size T before and after being delayed.

As a result of this embedding scheme, the expected value of the aggregate centroid, that is, the average of the arrival time of the packets modulo length of the interval T in either the intervals A_i (when watermark bit is 1) or B_i (when watermark bit is 0) corresponding to bit i, is increased by $a/2$. The difference between the aggregate centroid of A_i and B_i now will be $a/2$ when watermark bit is 1 or $-a/2$ when watermark bit is 0.

6.3.2.1 Detector. The detector checks for the existence of the watermark bits. The check on watermark bit i is performed by testing whether the expected difference of the aggregate centroid of packet arrival times in the intervals A_i and B_i is closer to $a/2$, or $-a/2$. If it is closer to $a/2$, then the watermark bit is decoded as 1, and if it is closer to $-a/2$, the bit is declared 0. By focusing on the arrival times of many intervals (r of them for each bit of watermark) rather than individual packet timings, the ICBW approach is robust to repacketization, insertion of chaff, and mixing of data flows. Network jitter can shift packets from one interval into another, but the suggested parameters for a and T (350 and 500 ms, respectively) are large enough, so a few packets will be affected.

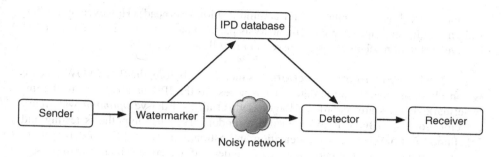

Figure 6.8. Model of RAINBOW network flow watermarking system.

6.3.3 RAINBOW System

RAINBOW [3] is a robust and invisible non-blind watermark that uses the timing patterns of network flows for watermarking. RAINBOW is the first *non-blind* flow watermarking scheme. The authors use a non-blind approach to be able to provide higher invisibility guarantees compared to previous blind designs.

The RAINBOW watermark embedding process is shown in Figure 6.8. Suppose that a flow with the packet timing information $\{t_i^u | i = 1, \ldots, n+1\}$ enters the border router where it is to be watermarked (we use the superscript u to denote an "unwatermarked" flow). Before embedding the watermark, the interpacket delays (IPD) of the flow, $\tau_i^u = t_{i+1}^u - t_i^u$, are recorded in an IPD database, which is accessible by the watermark detector. The watermark is subsequently embedded by delaying the packets by an amount such that the IPD of the ith watermarked packet is $\tau_i^w = \tau_i^u + w_i$. The watermark components $\{w_i\}_{i=1}^n$ take values $\pm a$ with equal probability. The value a is chosen to be small enough so that the artificial jitter caused by watermark embedding is invisible to ordinary users and attackers who aim at detecting any watermark presence.

In order to apply watermark delays on the flow, output packet t_i is delayed by $w_0 + \sum_{j=1}^{i-1} w_i$, where w_0 is the initial delay applied to the first packet. This results in $\tau_i^w = \tau_i^u + w_i$, as desired. Since a watermarker cannot delay a packet for a negative amount of time, w_0 must be chosen large enough to prevent this from happening. Since the sequence w_i is generated from a random seed, the watermarker can calculate all of the partial sums $\sum_{j=1}^{i-1} w_i$ in advance and adjust w_0 accordingly. If a particular random seed requires a very large initial delay w_0, a different seed can be chosen.

Once the watermarked flow traverses the network, it accumulates extra delays, that is, the noise from the communication network. Let d_i be the additional delay that the ith packet accumulates by the time it reaches a watermark detector (i.e., the packet is received at the detector at the time $t_i^r = t_i^w + d_i$). The IPD values at the detector are then

$$\tau_i^r = t_{i+1}^r - t_i^r = \tau_i^u + w_i + \delta_i, \tag{6.1}$$

where $\delta_i = d_{i+1} - d_i$ is the jitter present in the network.

As mentioned before, RAINBOW uses a non-blind flow watermarking scheme. Non-blind watermarking inherits similar scalability issues from the passive correlation schemes (e.g., high computation and communication overhead). However, non-blind watermarking can improve the traffic analysis performance compared to the traditional passive correlation schemes, as discussed in the following sections.

6.3.3.1 *Watermark Detection.*

As mentioned before, the RAINBOW scheme is non-blind and therefore the detector has access to the IPD database where the un-watermarked flows are recorded. Houmansadr et al. [36] derive *optimum* watermark detectors for RAINBOW in different threat models. They use hypothesis testing and Likelihood Ratio Tests from the detection and estimation theory [37] to derive these detectors, and compare the performance of the derived detection schemes with passive flow correlation schemes. The comparisons demonstrate that the non-blind watermark of RAINBOW outperforms passive correlation schemes in different traffic models; this confirms what is intuitively expected from information theory, since a non-blind watermark detector has access to more information (e.g., the watermark and the IPDs), compared to a passive correlation scheme, which only has access to the IPDs. They also show that the RAINBOW detector is reliable in different models, while the optimum passive correlation schemes fail in some scenarios. Note that this encourages the use of non-blind watermarks over passive correlation schemes, despite the fact that both have similar scalability limitations.

6.3.4 SWIRL System

SWIRL [34] is a blind flow watermark that uses the timing domain for marking flows. It is an interval-based timing watermark; therefore, it considers the flow as a collection of intervals of length T, with an initial offset o; that is, the ith interval includes packets during time period $[o + iT, o + (i + 1)T)$. In order to resist an earlier attack on interval-based watermarks, the MFA [38] attack, SWIRL proposes to perform interval-based marking in a flow-dependent manner. SWIRL's approach to flow-dependent marking is presented first, followed by the description of the overall SWIRL scheme.

6.3.4.1 *Flow-Dependent Marking.*

To perform flow-dependent marking, SWIRL selects two intervals: a *base* and a *mark* interval. The positions of these intervals will be fixed for all flows, but are otherwise arbitrary, with the restriction that the base interval must come earlier. During watermarking, SWIRL will use the base interval to decide which pattern to insert on the mark interval; the detector will correspondingly look for the pattern it computes using its version of the base interval.

The property of the base interval that SWIRL uses is the interval *centroid*, which is the average distance of the packets from the start of the interval. That is, if the interval i contains packets arriving at times $t_1, \ldots, t_n$, the centroid is

$$C = \frac{1}{n} \sum_{j=1}^{n} \left(t_j - (o + iT) \right). \tag{6.2}$$

To decide on the pattern to be used on the mark interval, the centroid is quantized to a symbol s in the range $[0, m)$ for some $m \in \mathbb{Z}_+$. Since the range of the centroid is

$[0, T)$, a simple approach would be to set $s = \lfloor mC/T \rfloor$. However, this would result in a nonuniform distribution for s, since a centroid is more likely to be in the center than at the interval. The actual distribution of centroids is heavily dependent on the rate of the flow as well as on the distribution of packet delays. In order to approximate a uniform distribution for s, we take the approach of using finer partitioning. Namely, we set

$$s = \lfloor qmC/T \rfloor \bmod m \qquad (6.3)$$

for $q > 1$. The quantization multiplier q helps smooth out the distribution: a larger value of q makes the value of s more sensitive to the value of C, that is, a smaller change in C is needed for s to change its value. To see this effect better, note that $C \in [0\ T]$, based on the definition of C in (6.2). The way (6.3) works is that for any q, it divides the range of C (i.e., $[0\ T]$) into $\lceil qm \rceil$ subsequent, nonoverlapping subintervals, with equal length $\lfloor T/(mq) \rfloor$ (except for the very last subinterval). Let us assign a number in $\{1, \ldots, \lceil qm \rceil\}$ to each subinterval based on its order of appearance. The value of $s = k$ is returned by (6.3) if C appears in any of the intervals with numbers αk, where $\alpha \in \mathbb{N}$. We can see that for $q = 1$, there is only a large subinterval that results in $s = k$; however, for larger q, the values of $s = k$ result from a number of smaller subintervals in the range $[0\ T]$. As a result, for larger q, the value of each symbol comes from different parts of C's range, which smoothes out the distribution of s.

The value s is then used to transform the mark interval. The watermarker first subdivides the mark interval into r subintervals of length T/r each. The subintervals are then further subdivided into m slots each, with the slots numbered $0, \ldots, m - 1$ (see Figure 6.9). The watermarker selects a slot in each subinterval i by applying a permutation $\pi^{(i)}$ to s; each packet is then delayed such that it falls within a selected slot, possibly moving into the next subinterval. (Any packets at the end of the interval past the last selected slot are not delayed.) This produces a distinct pattern in the mark interval; see Figure 6.10 for an illustration. SWIRL uses a distinct permutation for each subinterval to avoid a detectable periodic pattern. The permutations $\pi^{(0)}, \ldots, \pi^{(r-1)}$ are part of the secret watermark key.

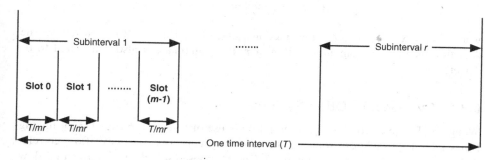

Figure 6.9. Slot numbering in the SWIRL scheme (Reproduced from [14] with permission of Springer.)

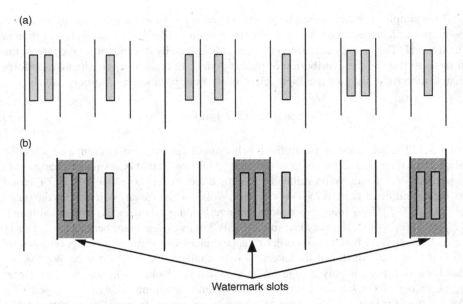

Figure 6.10. Delaying packets to insert a watermark by SWIRL. (Reproduced from [34] with permission of Springer.)

6.3.4.2 Detection. To detect the watermark presence, the detector analyzes packets in the base interval to compute the centroid $\hat{C}$. It then derives $\hat{s}$ from $\hat{C}$ using (6.3). It then counts the fraction of packets in the mark interval that are in the correct slot ($\pi^{(i)}(\hat{s})$). If this ratio ρ is greater than a packet threshold τ, then the watermark is considered to be detected successfully.

Note that the centroid of the interval may have shifted due to network noise. Consider an alternative quantization of it, $\hat{s}'$, to be the next nearest quantization to $\hat{C}$:

$$\hat{s}' = \begin{cases} \lceil mq\hat{C}/T \rceil \bmod m, & \text{if } \{mq\hat{C}/T\} \geq 0.5, \\ \lfloor mq\hat{C}/T \rfloor - 1 \bmod m, & \text{otherwise,} \end{cases} \tag{6.4}$$

where $\{x\} = x - \lfloor x \rfloor$ denotes the fractional part of x. The detector then repeats the detection using $\hat{s}'$ to compute ρ'. If $\rho' > \tau$, the watermark is also considered to be detected.

6.3.5 Overview of Other Systems

Wang and Reeves were the first to propose the use of flow watermarks [1] for linking flows in order to address some of the efficiency concerns of passive flow correlation schemes, discussed in Section 6.1.3. Their proposed scheme is timing-based flow watermarking that manipulates IPDs for marking flows. The IPD-based watermark of [1] was later shown to be detectable [39], in addition to being susceptible to packet modification, for example, packet additions/removals. To be robust against repacketization, Pyun et al. proposed an interval-based watermarking scheme in [2] by delaying packets of some intervals based on the watermark value. Dealing with intervals instead of packets themselves makes the detection scheme robust to packet addition/removal. Some of the schemes target anonymous communication [5,6] rather than stepping stones as the application area, but the techniques for both are comparable. In [5] Wang et al. use an interval-based watermarking scheme similar to [2] aiming to compromise anonymity of Anonymizer [24]. Yu et al. suggested another interval-based watermark [6] by changing the packet rates of flow intervals using direct sequence spread spectrum (DSSS). More recently, Jia et al. have shown how this watermark is detectable exploiting statistical attacks against DSSS [40]. Another watermark focused on anonymous networks was proposed by Wang et al. [4], which tracks anonymous peer-to-peer VoIP calls over the Internet.

Ling et al. [35] proposed a Tor-specific flow watermark that works by embedding a secret watermark sequence in a Tor flow through making modifications to cell [41] counters. More specifically, a compromised/malicious Tor relay manipulates the number of Tor cells being sent together on a Tor circuit so that each single cell denotes a "0" bit and each triple cell denotes a "1" bit. Houmansadr and Borisov [42] demonstrate that the use of communication codes (e.g., Repeat Accumulate codes) in the design of watermark signals can improve the detection performance of flow watermarking systems.

6.4 WATERMARKING VERSUS FINGERPRINTING

In this section, we introduce a variant of flow watermarking, called *flow fingerprinting*. Flow fingerprinting is an underexplored type of active traffic analysis that was first proposed by Houmansadr and Borisov [43] for linking network flows. Information, theoretically, flow watermarking aims at conveying a *single bit* of information whereas flow fingerprinting tries to reliably send *multiple bits* of information; hence, it is a more challenging problem. Such additional bits help a fingerprinter deliver extra information in addition to the existence of the tag, such as the network origin of the flow and the identity of the fingerprinting entity.

Similar to flow watermarking, flow fingerprinting can be used in any networking application that meets the conditions described in Section 6.2. In the following, we compare the use of fingerprinting and watermarking for two common applications of active traffic analysis.

6.4.1 Compromising Anonymity Systems

An anonymity system like Tor [20] maps a number of ingress flows to a number of egress flows while hiding the relationships between these flows. The goal of an attacker, then, is to link an incoming flow to its outgoing flow (or vice versa). Previous research [4,5] has suggested the use of flow watermarks for performing this attack. To do so, an attacker tags the flows entering the anonymity network and watches output flows for the inserted watermark. Such an attack can be performed in two manners, *targeted* and *nontargeted*. As we discuss in the following, flow watermarking is able to conduct only the targeted form of this attack, whereas conducting the nontargeted attack requires flow fingerprinting.

1. *Targeted attack* Consider a malicious website who aims at identifying its visitors, even if they use an anonymity system like Tor for their connections (see Figure 6.11a). To do so, the malicious website inserts a tag on every network connection between itself and the target anonymizing system. An accomplice who can eavesdrop on a link to the anonymity system (e.g., a malicious Tor entry node, or an can identify the visitors of the malicious website by looking for the inserted tag. Note that, in this case, the malicious website suffices to insert *the same tag* on any flow that it tags, that is, it inserts a watermark. This is because the accomplice only needs to check for the existence of the tag, but not its value.

2. *Nontargeted attack* Now consider a different scenario in which two (or more) compromised/malicious Tor [20] nodes try to deanonymize Tor's connections (see Figure 11b). We argue that this needs to be performed through flow fingerprinting, but not flow watermarking. Suppose that the malicious nodes *A* and *B*

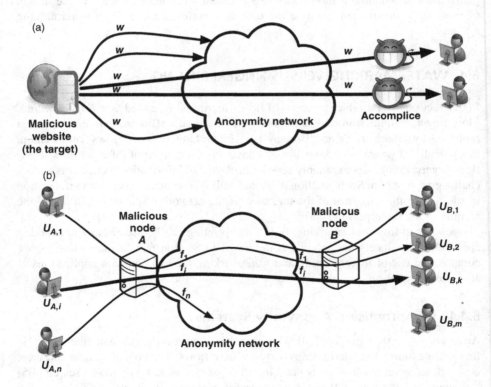

Figure 6.11. Targeted (a) and nontargeted (b) attacks on an anonymous network.

intercept connections from n and m distinct users, respectively. The use of a flow watermark in this scenario means that node A will insert the *same* watermark tag on every flow. Therefore, if B detects the watermark on the traffic of one of its m users, namely, $U_{B,k}$, the colluding attackers (B and A) can only infer that $U_{B,k}$ is communicating with *one of* the n users watermarked by A, but they will not be able to tell *with which one of them*. Now suppose that flow fingerprinting is used by the attackers in this scenario. In this case, for each of the n users A intercepts, A will insert a different tag tailored to that user (i.e., a fingerprint), for example, it inserts the fingerprint f_i on the traffic of user $U_{A,i}$. Therefore, once B observes that the traffic to one of its users, $U_{B,k}$, contains the fingerprint f_i, the colluding attackers A and B will be able to infer that users $U_{A,i}$ and $U_{B,k}$ are communicating through the anonymity system.

6.4.2 Stepping Stone Detection

As introduced earlier, a stepping stone is a host that is used to relay attack traffic to its victim destination, in order to hide the true origin of the attack. To defend, an enterprise network should be able to identify the ingress flows that are linked (correlated) with some egress flow. The situation is therefore very similar to an anonymous communication system, with n flows entering the enterprise and m flows leaving. There are two objectives for active traffic analysis in this case, as described in the following; the first one is achievable using flow watermarking, while the second one requires flow fingerprinting.

1. *Detecting relayed flows* As previous research [2,3] suggests, flow watermarks can be used to detect the network flows relayed through stepping stones. Suppose that the enterprise network consists of two border routers A and B. To link the relayed flows, the border router A inserts a watermark tag w in all flows that enter the enterprise network. On the other side, the border router B inspects all egress network flows, looking for the watermark w. Suppose that A intercepts n flows and B intercepts m network flows at a given time. If B detects that a network flow $F_{B,k}$ is carrying the watermark tag w, the security officer of the enterprise network infers that $F_{B,k}$ is traffic relayed through the enterprise. However, the security officer cannot tell which of the n flows observed by A is the source of $F_{B,k}$, since A inserts the same watermark tag on all intercepted flows.

2. *Detecting relayed flows and their origins* Flow fingerprinting can be used to not only detect the relayed flows but also identify their sources. Consider the case in which the border router A inserts different tags (i.e., fingerprints) on each of the n intercepted flows (i.e., the fingerprint f_i is inserted into the ith flow, $F_{A,i}$). Now, suppose that B detects the fingerprint f_i on the network flow $F_{B,k}$. In this case, the security officer infers two facts: First, $F_{B,k}$ is relayed traffic and, second, the source of this relay traffic is the network flow $F_{A,i}$. A watermark, however, is not able to identify the source of the relayed traffic.

6.5 CHALLENGES OF FLOW WATERMARKING

A flow watermarking system's effectiveness is evaluated based on the features introduced in Section 6.1.4, that is, robustness, invisibility, and detection accuracy. In this section, we describe various issues that may impact a flow watermark's robustness, invisibility, and detection accuracy.

6.5.1 Natural Network Perturbations

After being watermarked, a network flow usually traverses a *noisy communication channel* before getting intercepted and evaluated by a watermark detector. For instance, in the application scenario of stepping stone detection, a watermarked flow has to pass through a noisy computer network, for example, a corporate network, before reaching a detector. A noisy communication channel may perturb network flows in different ways, for example, timing delays/jitter on packets, packet drops, packet retransmissions, and repacketization. For a flow watermark to be effective, it should be robust to such natural network perturbations.

Different designs of flow watermarks have taken various approaches toward providing such robustness.

6.5.1.1 Redundant Mark Insertion. The use of redundancy is prevalent in the design of flow watermarks to improve robustness. For instance, the ICBW scheme introduced in Section 6.3.2 repeats a watermark's timing pattern multiple times on a single flow to improve resilience to network perturbation.

6.5.1.2 Interval-Based Mark Insertion. Several proposals [2,5,6,34] embed the watermark inside timing intervals of network flows. This approach, that is, interval-based watermarking, improves watermark robustness against various kinds of perturbation such as packet drops, packet retransmissions, and repacketization.

6.5.1.3 Sliding Window Correlation. The RAINBOW watermark, described in Section 6.3.3, uses sliding window correlation to protect the watermark against dropped/repacketized packets.

6.5.2 Adversarial Countermeasures

Please refer to Chapter 8 for discussions on passive and active countermeasures against network flow watermarking systems by adversaries.

6.6 SUMMARY

Network flow watermarking is a particular type of network steganography that hides information into the communication patterns of traffic, for example, packet timings and packet sizes. This chapter introduced network flow watermarks, and reviewed its application to link network flows in different scenarios.

REFERENCES

1. X. Wang and D. S. Reeves. Robust correlation of encrypted attack traffic through stepping stones by manipulation of interpacket delays. In V. Atluri, editor, *ACM Conference on Computer and Communications Security*, pp. 20–29. ACM, New York, NY, 2003.

2. Y. Pyun, Y. Park, X. Wang, D. S. Reeves, and P. Ning. Tracing traffic through intermediate hosts that repacketize flows. In G. Kesidis, E. Modiano, and R. Srikant, editors, *IEEE Conference on Computer Communications (INFOCOM)*, pp. 634–642. IEEE, May 2007.

3. A. Houmansadr, N. Kiyavash, and N. Borisov. RAINBOW: a robust and invisible non-blind watermark for network flows. In *Network and Distributed System Security Symposium*, February 2009.

4. X. Wang, S. Chen, and S. Jajodia. Tracking anonymous peer-to-peer VoIP calls on the Internet. In C. Meadows, editor, *ACM Conference on Computer and Communications Security*, pp. 81–91. ACM, New York, NY, November 2005.

5. X. Wang, S. Chen, and S. Jajodia. Network flow watermarking attack on low-latency anonymous communication systems. In *IEEE Symposium on Security and Privacy*, pp. 116–130, 2007.

6. W. Yu, X. Fu, S. Graham, D.Xuan, and W. Zhao. DSSS-based flow marking technique for invisible traceback. In *IEEE Symposium on Security and Privacy*, pp. 18–32, 2007.

7. K. Yoda and H. Etoh. Finding a connection chain for tracing intruders. In F. Cuppens, Y. Deswarte, D. Gollmann, and M. Waidner, editors, *European Symposium on Research in Computer Security*, Vol. 1895 of *Lecture Notes in Computer Science*, pp. 191–205. Springer, 2000.

8. Y. Zhang and V. Paxson. Detecting stepping stones. In S. Bellovin and G. Rose, editors, *USENIX Security Symposium*, pp. 171–184. USENIX Association, Berkeley, CA, 2000.

9. X. Wang, D. Reeves, and S. F. Wu. Inter-packet delay based correlation for tracing encrypted connections through stepping stones. In D. Gollmann, G. Karjoth, and M. Waidner, editors, *European Symposium on Research in Computer Security*, Vol. 2502 of *Lecture Notes in Computer Science*, pp. 244–263. Springer, 2002.

10. D. Donoho, A. Flesia, U. Shankar, V. Paxson, J. Coit, and S. Staniford. Multiscale stepping-stone detection: detecting pairs of jittered interactive streams by exploiting maximum tolerable delay. In A. Wespi, G. Vigna, and L. Deri, editors, *International Symposium on Recent Advances in Intrusion Detection*, Vol. 2516 of *Lecture Notes in Computer Science*, pp. 16–18. Springer, 2002.

11. A. Blum, D. X. Song, and S. Venkataraman. Detection of interactive stepping stones: algorithms and confidence bounds. In E. Jonsson, A. Valdes, and M. Almgren, editors, *International Symposium on Recent Advances in Intrusion Detection*, Vol. 3224 of *Lecture Notes in Computer Science*, pp. 258–277. Springer, 2004.

12. C. V. Wright, L. Ballard, F. Monrose, and G. M. Masson. Language identification of encrypted VoIP traffic: Alejandra y roberto or Alice and Bob? In *USENIX Security*, 2007.

13. M. Backes, G. Doychev, M. Dürmuth, and B. Köpf. Speaker recognition in encrypted voice streams. In *European Symposium on Research in Computer Security (ESORICS)*, pp. 508–523. Springer, 2010.

14. Y. Lu. *On Traffic Analysis Attacks to Encrypted VoIP Calls*. PhD thesis, Cleveland State University, 2009.

15. C. V. Wright, L. Ballard, S. E. Coull, F. Monrose, and G. M. Masson. Spot me if you can: uncovering spoken phrases in encrypted VoIP conversations. In *IEEE Symposium on Security and Privacy*, pp. 35–49, 2008.

16. A. M. White, A. R. Matthews, K. Z. Snow, and F. Monrose. Phonotactic reconstruction of encrypted VoIP conversations: Hookt on fon-iks. In *IEEE Symposium on Security and Privacy*, pp. 3–18, 2011.

17. S. Staniford-Chen and L. T. Heberlein. Holding intruders accountable on the Internet. In C. Meadows and J. McHugh, editors, *IEEE Symposium on Security and Privacy*, pp. 39–49. IEEE Computer Society Press, May 1995.

18. A. Houmansadr and N. Borisov. Botmosaic: collaborative network watermark for botnet detection. Technical Report arXiv:1203.1568v1, Computing Research Repository arXiv, March 2012.

19. D. Ramsbrock, X. Wang, and X. Jiang. A first step towards live botmaster traceback. In *Recent Advances in Intrusion Detection*, pp. 59–77, 2008.

20. R. Dingledine, N. Mathewson, and P. Syverson. Tor: the second-generation onion router. In M. Blaze, editor, *USENIX Security Symposium*. USENIX Association, Berkeley, CA, 2004.

21. G. Danezis, R. Dingledine, and N. Mathewson. Mixminion: design of a type III anonymous remailer protocol. In *IEEE Symposium on Security and Privacy*, pp. 2–15. IEEE Computer Society Press, Berkeley, CA, 2003.

22. T. Ylonen and C. Lonvick. The secure shell (SSH) protocol architecture. RFC 4251, January 2006.

23. P. Syverson, G. Tsudik, M. Reed, and C. Landwehr. Towards an analysis of onion routing security. In *Proceedings of Designing Privacy Enhancing Technologies, Vol. 2009 Of Lecture Notes in Computer Science*, pp. 96–114. Springer, Berlin, 2009.

24. J. Boyan. The anonymizer: protecting user privacy on the web. *Computer-Mediated Communication Magazine*, 4(9), September 1997.

25. P. Syverson, G. Tsudik, M. Reed, and C. Landwehr. Towards an analysis of onion routing security. In H. Federrath, editor, *Proceedings of Designing Privacy Enhancing Technologies: Workshop on Design Issues in Anonymity and Unobservability*, pp. 96–114. Springer, July 2000.

26. A. Back, I. Goldberg, and A. Shostack. Freedom systems 2.1 security issues and analysis. White paper, Zero Knowledge Systems, Inc., May 2001.

27. A. Ramachandran and N. Feamster. Understanding the network-level behavior of spammers. In L. Rizzo, T. E. Anderson, and N. McKeown, editors, *SIGCOMM*, pp. 291–302. ACM, 2006.

28. L. Kharouni. SDBOT IRC botnet continues to make waves. White paper, Trend Micro Threat Research, December 2009.

29. J. Zhuge, T. Holz, X. Han, J. Guo, and W. Zou. Characterizing the IRC-based botnet phenomenon. Technical Report TR-2007-010, Department for Mathematics and Computer Science, University of Mannheim, 2007.

30. A. Houmansadr and N. Borisov. BotMosaic: collaborative network watermark for the detection of IRC-based botnets. *Journal of Systems and Software*, 86(3):707–715, 2013.

31. L. Spitzner. The Honeynet Project: trapping the hackers. *IEEE Security & Privacy Magazine*, 1(2):15–23, 2003.

32. N. Evans, R. Dingledine, and C. Grothoff. A practical congestion attack on Tor using long paths. In F. Monrose, editor, *USENIX Security Symposium*, pp. 33–50. USENIX, 2009.

33. S. Murdoch and G. Danezis. Low-cost traffic analysis of Tor. In V. Paxson and M. Waidner, editors, *IEEE Symposium on Security and Privacy*. IEEE Computer Society, May 2005.

34. A. Houmansadr and N. Borisov. SWIRL: a scalable watermark to detect correlated network flows. In *Proceedings of the Network and Distributed System Security Symposium (NDSS'11)*, 2011.

35. Z. Ling, J. Luo, W. Yu, X. Fu, D. Xuan, and W. Jia. A new cell counter based attack against Tor. In *ACM Conference on Computer and Communications Security (CCS)*, New York, NY, 2009.

36. A. Houmansadr, N. Kiyavash, and N. Borisov. Non-blind watermarking of network flows. *IEEE/ACM Transactions on Networking*, 22(4):1232–1244, 2014.

37. H. V. Poor. *An Introduction to Signal Detection and Estimation*. Springer, 1998.

38. N. Kiyavash, A. Houmansadr, and N. Borisov. Multi-flow attacks against network flow watermarking schemes. In P. van Oorschot, editor, *USENIX Security Symposium*. USENIX Association, Berkeley, CA, 2008.

39. P. Peng, P. Ning, and D. S. Reeves. On the secrecy of timing-based active watermarking trace-back techniques. In V. Paxson and B. Pfitzmann, editors, *IEEE Symposium on Security and Privacy*, pp. 334–349. IEEE Computer Society Press, May 2006.

40. W. Jia, F. P. Tso, Z. Ling, X. Fu, D. Xuan, and W. Yu. Blind detection of spread spectrum flow watermarks. In *INFOCOM*. IEEE, 2009.

41. R. Dingledine and N. Mathewson. Tor protocol specification. https://gitweb.torproject.org/torspec.git?a=blob_plain;hb=HEAD;f=tor-spec.txt.

42. A. Houmansadr and N. Borisov. Towards improving network flow watermarking using the repeat-accumulate codes. In *36th IEEE International Conference on Acoustics, Speech and Signal Processing (ICASSP)*, 2011.

43. A. Houmansadr and N. Borisov. The need for flow fingerprints to link correlated network flows. In *Proceedings of the 13th Privacy Enhancing Technologies Symposium (PETS)*, 2013.

7

EXAMPLES OF INFORMATION HIDING METHODS FOR POPULAR INTERNET SERVICES

Three may keep a secret, if two of them are dead.

—Benjamin Franklin, Poor Richard's Almanack

This chapter provides some examples of current information hiding techniques for communication protocols of popular Internet applications and services. Using application-layer protocols as cover traffic has several advantages. The high complexity of many application protocols means there are many opportunities for embedding covert channels, the steganographic bandwidth is often higher than for channels at a lower layer, and channels at application layer can exploit the properties of the whole underlying protocol stack, such as reliable data transport.

First, hiding concepts for IP-based telephony are presented together with *transcoding*, a method relying on data compression. We moreover show how data can be hidden in popular P2P services, namely Skype and BitTorrent, and how steganography can be realized for mobile devices, such as smartphones by exploiting services like Siri. We continue with the topic of information hiding in new network protocols (IPv6 and SCTP) and in Wi-Fi networks. Afterwards, we will discuss covert channels in online multiplayer games, specifically first person shooter games. The chapter ends with a

Information Hiding in Communication Networks: Fundamentals, Mechanisms, Applications, and Countermeasures,
First Edition. Wojciech Mazurczyk, Steffen Wendzel, Sebastian Zander, Amir Houmansadr, and Krzysztof Szczypiorski.
© 2016 by The Institute of Electrical and Electronics Engineers, Inc. Published 2016 by John Wiley & Sons, Inc.

discussion of methods to transfer covert information over online social networks and the exploitation of the Internet of Things for steganographic purposes.

7.1 IP TELEPHONY: BASICS AND INFORMATION HIDING CONCEPTS

IP telephony or Voice over IP (VoIP) is a real-time service that enables users to make phone calls through IP data networks. It is one of the most important services of IP-based networks that has changed the entire telecommunications landscape. Typically, an IP telephony connection consists of two phases in which certain types of traffic are exchanged between the parties participating in a call: signaling and conversation phases. During the first phase certain signaling protocol messages, for example, Session Initiation Protocol (SIP) messages [1] (often together with the Session Description Protocol (SDP) that describes multimedia communication session) are exchanged between the caller and callee. These messages are intended to set up and negotiate the connection parameters between the call parties. Then, during the second phase two audio streams are sent bidirectionally. The Real-Time Transport Protocol (RTP) [2] is most often utilized for voice data transport and thus packets that carry the voice payload are called RTP packets. Consecutive RTP packets form an RTP stream.

IP telephony is attracting increasing attention of the steganography research community because of its features that makes it a perfect carrier for steganographic purposes:

- *Popularity.* The more common carrier the better so its usage will not raise suspicions, that is, it will not be considered as an anomaly itself.
- *The large volume of VoIP data.* The more frequent the presence and utilization of such carriers in networks, the better their masking capacity, as hidden communications can pass unnoticed amongst the bulk of exchanged data.
- *Potentially high steganographic bandwidth.* For example, during the conversation phase of a G.711-based call, each RTP packet carries 20 ms of voice; in this case the RTP stream rate is 50 packets per second. Thus, even by only hiding 1 bit in every RTP packet, it is possible to achieve reasonably high steganographic bandwidth of 50 bit/s.
- *Combined use of a variety of protocols.* Many opportunities for hiding information arise from the different layers of the TCP/IP stack for VoIP. Hidden communication can be enabled by employing steganographic methods applied to the users' voice that is carried inside the RTP packets' payload, by utilizing so called well-known digital media steganography, or by utilizing VoIP protocols as a steganographic carrier. This makes VoIP a multi-dimensional carrier.
- *Real-time service constraints.* This induces strict requirements for steganographic methods and their detection (steganalysis) counterparts. However, at the same time it creates new opportunities for steganography (e.g., utilization of excessively delayed packets that are discarded by the receiver without processing, because they cannot be considered for voice reconstruction).

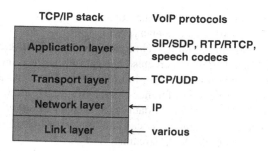

<u>Figure 7.1.</u> The VoIP stack and protocols. (Reproduced from [4] with permission of IEEE.)

- *Dynamics of VoIP calls.* IP telephony conversations are dynamic and of variable length, which make VoIP-based steganography even harder to detect.

Presently, steganographic methods that can be utilized for IP telephony are jointly described by the term VoIP steganography or steganophony [3]. These terms pertain to the techniques of hiding information in any layer of the TCP/IP protocol stack (Figure 7.1), including techniques applied to the speech codecs, or those that utilize the speech itself.

IP telephony as a hidden data carrier can be considered a fairly recent discovery. However, existing VoIP steganographic methods stem from two distinct research origins.

The first is the well-established image, audio and video file steganography (also called digital media steganography) [5], which has given rise to methods that target the digital representation of the transmitted voice as the carrier for hidden data. The second sphere of influence is the so-called covert channels, created in different network protocols [6] that target specific VoIP protocol fields (e.g., fields in the SIP signalling protocol, RTP transport protocol, or RTCP control protocol), or these protocols' behavior.

In the following subsections, we analyze VoIP steganography methods taking into account the second research origins mentioned above.

7.1.1 Steganographic Methods Applied to VoIP-specific Protocols

Utilization of the VoIP-specific protocols as a secret data carrier was first presented by Wang et al. [7]. The authors proposed embedding a 24-bit watermark into the encrypted stream (e.g., Skype call) to track its propagation through the network, thus providing its deanonymization. The watermark is inserted by modifying the inter-packet delay for selected packets in the VoIP stream. Wang et al. demonstrated that depending on the watermark parameters chosen, they are able to achieve a 99% true positive and 0% false positive rate while maintaining good robustness and undetectability. However, they achieved a steganographic bandwidth of only about 0.3 bit/s which is enough for watermarking purposes but rather low for clandestine communication.

Shah *et al.* [8] investigated the use of injected jitter into VoIP packets to create a covert channel. The channel is intended to exfiltrate users' keyboard activity, for example, authentication credentials and the authors also proved that such an attack is feasible even when the VoIP stream is encrypted.

A new kind of information hiding technique called interference channel was introduced by Shah and Blaze [9], which creates external interference on a shared communications medium (e.g., wireless network) in order to send hidden data. They describe an implementation of a wireless interference channel for 802.11 networks that is able to covertly transfer secret information over data streams (with a rather low steganographic bandwidth of 1 bit per 2.5 seconds of the call) and that is proven to be especially well suited for UDP-based traffic like VoIP streams.

Mazurczyk and Kotulski [10] proposed the use of steganography and digital watermarking to embed additional information into RTP traffic to provide origin authentication and content integrity. The necessary information was embedded into unused fields in the IP, UDP and RTP protocol headers, and also into the transmitted voice. The authors later further enhanced their scheme [11] by also incorporating a Real-time Transport Control Protocol (RTCP) functionality without the need to use a separate protocol, thus reducing the bandwidth utilized by VoIP connection.

A broader view on network steganography methods that can be applied to VoIP, to its signalling protocol, SIP with SDP [12], and to its RTP streams (also with RTCP) [13] was presented by Mazurczyk and Szczypiorski. They discovered that a combination of information hiding solutions provides a capacity to covertly transfer about 2 kbit/s during the signalling phase of a connection and about 2.5 kbit/s during the conversation phase. In 2010 Lloyd [14] extended Mazurczyk and Szczypiorski work [12] by introducing further steganographic methods for SIP and SDP protocols and by performing real-life experiments to verify whether they are feasible.

Bai *et al.*, [15] proposed a covert channel based on the jitter field of the RTCP header. The method has two stages: first, statistics of the jitter field value in the current network are calculated. Then, the secret message is modulated into the jitter field according to the previously calculated parameters. This guarantees that the characteristics of the covert channel are similar to those of the overt channel.

Forbes [16] proposed an RTP-based steganographic method that modifies the timestamp value of the RTP header to send steganograms. The method's theoretical maximum steganographic bandwidth is 350 bit/s.

Wieser and Röning [17] performed real-life experiments with VoIP steganography on a real-life device, SBC (Session Border Controller), that was acting as a gatekeeper. The authors tried to establish whether an SBC implements some countermeasures against information hiding techniques based on SIP and RTP protocols. The results showed that it was possible to covertly send data with a high steganographic bandwidth of even up to 569 kB/s and still remain undetectable.

Huang *et al.* [18] described how to provide efficient cryptographic key distribution in a VoIP environment for covert communication. Their proposed steganographic method is based on the utilization of the NTP timestamp field of the RTCP's SR (Sender Report) as a hidden data carrier, and offered steganographic bandwidth of 54 bit/s with good undetectability.

Mazurczyk et al. [19] introduced TranSteg, a steganographic method that relies on the compression of the overt data in a payload field of RTP packets in order to make space for secret data. For a chosen voice stream, TranSteg finds a codec that provides a similar voice quality as the original codec, but produces a smaller voice payload size. Then, the voice stream is transcoded: the original voice payload size is intentionally unaltered and the change of the codec is not indicated. Instead, after placing the transcoded voice payload, the remaining space is filled with hidden data. The resulting steganographic bandwidth that was obtained using a proof-of-concept implementation was 32 kbit/s while introducing delays lower than 1 ms, and still retaining good voice quality. The work was further extended by analyzing the influence of speech codec selection on TranSteg efficiency [20]. An interesting finding was that if the codec pair G.711/G.711.0 is utilized TranSteg introduces no steganographic cost and it offers a remarkably high steganographic bandwidth, on average about 31 kbit/s.

Tian et al. [21] experimentally evaluated the steganographic bandwidth and undetectability of two VoIP steganography methods proposed earlier by Huang et al. [18] (LSBs of the NTP Timestamp field of RTCP protocol) and by Forbes [16] (LSBs of Timestamp field of RTP protocol). They utilized the Windows Live Messenger voice conversations system and proved that the first approach achieves a steganographic bandwidth of 335 bit/s and the the second approach still reaches a bandwidth of 5.1 bit/s. The latter method is harder to detect.

Mazurczyk and Szczypiorski [13] introduced a novel, steganographic method called LACK (Lost Audio Packets Steganography) that is described in details in Section 3.3.

7.1.2 Transcoding Steganography as VoIP Steganography Example

As an example of VoIP steganography, Transcoding Steganography (TranSteg) [19] will be presented in detail. TranSteg's main application is IP telephony, but it can also be used for other applications or services (like video streaming) where a possibility exists to efficiently compress (in a lossy or lossless manner) the overt data.

The typical approach to steganography is to compress the covert data in order to limit its size (it is reasonable in the context of a limited steganographic bandwidth). In TranSteg, compression of the overt data is used to make space for the steganogram—the concept is similar to the invertible authentication watermark that was first proposed by Fridrich et al. [22] for JPEG images. TranSteg is based on the idea of voice data transcoding (lossy compression) from a higher bit rate codec (and thus greater voice payload size) to a lower bit rate codec (with smaller voice payload size) with the least possible degradation in voice quality.

Generally, Transcoding Steganography works as follows (Figure 7.2). First, RTP packets that carry the caller's voice are analyzed and the codec originally utilized for voice encoding (hereinafter referred as the overt codec) is identified by inspecting the PT (Payload Type) field in the RTP header (Figure 7.2, 1). Typically, if (nonsteganographic) transcoding is realized, then the original voice frames are encoded using a different voice codec and, as a result, a smaller voice frame is achieved (Figure 7.2, 2). However, TranSteg selects a so called covert codec for the originally utilized overt codec. The covert codec should yield a comparable voice quality but should result in a smaller voice payload size. The voice stream is transcoded (using the covert codec) to a smaller size and placed into the original payload field. However, original payload size and the

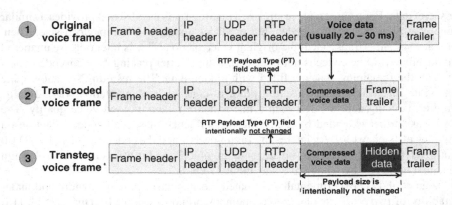

Figure 7.2. A frame carrying a speech payload encoded with an overt codec (1), transcoded (2), and encoded with a covert codec (3). (Reproduced from [19] with permission of IEEE.)

indicator of the codec type (in the PT field) are left unchanged. Thus, the remaining free space can be filled with secret information (Figure 7.2, 3). Note that the secret data can be put after the voice data or interleaved with the voice samples in a predetermined way.

The performance of TranSteg depends, most notably, on the characteristics of the two codecs: the overt codec and the covert codec. It is worth noting that, depending on the hidden communication scenario, TranSteg may or may not be able to influence the choice of overt codec (as it depends on whether steganographer is controlling an endpoint or not). It is assumed that it is always possible to find a covert codec for a given overt one. However, it must be noted that for very low bit rate codecs, the steganographic bandwidth is limited. Ideally, the covert codec should:

- not degrade the user voice quality considerably (due to the transcoding operation and the introduced delays), when compared to the quality of the overt codec; and
- provide the smallest achievable voice payload size as this results in the most free space in an RTP packet to insert secret data.

If there is a possibility to influence the overt codec, in an ideal situation it should

- result in the largest possible voice payload size to provide, together with the covert codec, the highest achievable steganographic bandwidth, and
- be used commonly to not raise suspicion.

Taking the above into account, TranSteg's steganographic bandwidth (S_B) can be expressed as

$$S_B = (P_O - P_C)PN_S [bit/s]. \tag{7.1}$$

P_O denotes the overt codec's payload size, P_C is the covert codec's payload size, and PN_S describes the number of RTP packets sent during one second.

Transcoding Steganography can be applied in four different communication scenarios (Figure 2.12). The first one is typically the most desired and common (S1 in Figure 2.12): the secret sender (SS) and the secret receiver (SR) setup an IP telephony call while exchanging secret messages (in an end-to-end manner). The hidden data path is identical to the conversation path. In the remaining scenarios, only a fragment of the end-to-end path of the VoIP call is utilized for hidden communication purposes (S2–S4 in Figure 2.12). Therefore, in principle, the overt sender or (and) the overt receiver are unaware of the steganographic data exchange. The application of TranSteg for VoIP service enables to transfer secret data while, simultaneously, still preserving users' conversation. As emphasized previously, it is especially vital for S2–S4.

Here we focus on TranSteg applied in scenario S4 (intermediary devices utilize third-party VoIP call for covert communication purpose) as this is the worst case scenario in terms of the speech quality, because it requires triple transcoding and two transcodings are caused by the TranSteg functioning alone. For other scenarios (S1–S3), we would avoid one or even two transcodings (in scenario S1); hence, the negative impact on voice quality would be significantly lower.

In S4, we assume that the secret sender and the secret receiver can capture and inspect all the voice packets transmitted between the caller and the callee (Figure 7.3). Since in scenario S4 neither SS nor SR are directly involved in the overt VoIP call, it is more difficult to spot the hidden data exchange compared to the other scenarios (1–3). Therefore, the secret data is transmitted only across the fragment of the communication path—thus, it never reaches the endpoints.

In S4, both hidden parties behave similarly. They must transcode the data: the sender must convert the speech payload from the overt to the covert codec and the secret receiver must conduct the reverse process. In the scenario S4, the secret sender behaves as follows:

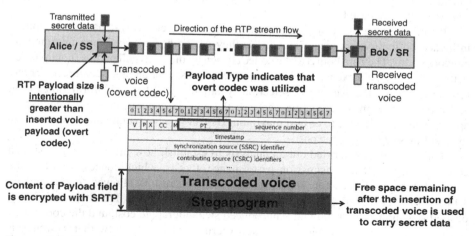

Figure 7.3. The TranSteg scenario S4 (SS–Secret Sender; SR–Secret Receiver). (Reproduced from [19] with permission of IEEE.)

- *Step 1:* It captures the incoming voice stream and transcodes the speech encoded with the overt codec to the covert codec.
- *Step 2:* The transcoded speech payload is inserted again in a voice packet. The RTP packet's header is not modified.
- *Step 3:* The resulting free space in the voice payload is populated with the secret data bits (and thus the rest of the original speech payload is erased).
- *Step4:* Lower layer protocol checksums are recalculated (e.g., the CRC at the link layer and the IP and UDP checksums).
- *Step5:* Encapsulated voice packets are then transmitted to the secret receiver in the modified frames.

When the changed voice stream reaches the secret receiver, it executes the following steps:

- *Step1:* It analyzes the voice payload and extracts the secret data bits from the consecutive RTP packets.
- *Step2:* The speech payload is then retranscoded (from the covert to the overt codec) and inserted in the payload of the packets. It overwrites the secret data bits with user speech. The RTP packet's header is not modified.
- *Step3:* Lower layer protocols' checksums are recalculated.
- *Step4:* The resulting voice packets (with the same voice payload type as the original voice packets sent) are then transmitted to the original receiver (callee).

The most important benefit of scenario S4 is its potential ability to utilize aggregated IP telephony traffic for steganographic purposes. If both secret sender and secret receiver can tap into more than one IP telephone call then the potential steganographic bandwidth can be greatly increased.

It must be noted that the secret sender and the secret receiver have only limited influence on the overt codec selection, as they are both placed at some intermediate network nodes. They are bound to the codec chosen by the overt call parties (not involved in the steganographic exchange). However, the secret sender and the secret receiver can possibly influence the overt codec selection by tampering exchanged information during codec negotiation in the signaling phase of the call. In general, they must choose an appropriate covert codec to ensure maximum steganographic bandwidth while limiting the resulting steganographic cost.

Janicki et al. [20] carried out an analysis to establish how the selection of speech codecs affects hidden transmission performance, that is, which codecs would be the most advantageous ones for TranSteg. The obtained results allowed to recommend 10 pairs of overt/covert codecs that can be used effectively for TranSteg considering desired steganographic bandwidth, introduced steganographic cost, and the codec used in the overt transmission. Moreover, the experimental results show that TranSteg is most effective when G.711 is used for the overt transmission. This is caused by several factors:

- G.711 has a high bitrate, so there is more potential space for hidden data;
- G.711 offers a high speech quality, thus a decrease in MOS is less noticeable;
- G.711 performs well if transcoded more than once, which we think is due to the fact that G.711 is a waveform codec; that is, it preserves the waveform shape;
- while being a waveform codec, G.711 behaves well if further transcoded with other codecs, especially CELP-based ones (AMR, Speex in mode 7);
- and, simple bit analysis cannot detect the channel, as a G.711 payload contains no headers or flags, but only ''random'' speech samples.

Therefore, when it is possible to select an overt codec (e.g., by imposing it in codec negotiation in scenario S3), G.711 is the best option. By contrast, Speex in the low mode (mode 4) and G.723.1 as the overt codecs turned out to provide too poor overall quality when cascaded with any of the covert codecs, and therefore should be avoided by TranSteg.

Another potential approach to TranSteg steganalysis was proposed by Janicki et al. [20]. It was based on the detection of a mismatch between the declared codec (set in PT field) and the actual voice codec used for transcoding. It was found that some codecs are fairly easily detected using a simple analysis of the first byte of the payload. Therefore, if we want to use, for example, Speex codecs, we must undertake additional actions (such as bit randomization), in order to make detection less likely.

When experimenting with various combinations of overt and covert codecs, it was also observed that some codecs do not complement each other well. For example, Speex in mode 7 works significantly better (in terms of voice quality) with AMR than with GSM 06.10, even though the two TranSteg configurations result in similar steganographic bandwidths. A similar phenomenon has been observed in other domains, for example in speaker recognition from coded speech, in situations where there was a mismatch between the codec used in voice transmission and the codec used to create speakers' models [23].

The choice of a covert codec depends on an actual application, or more precisely, on whether priority is given to the higher steganographic bandwidth, better speech quality, or higher undetectability. Janicki et al. [20] recommended for TranSteg several pairs of overt/covert codecs. These pairs were grouped into three classes based on the steganographic cost and the recommendations about security of a given pair were provided. The pair G.711/G.711.0 is costless, that is, it introduces no steganographic cost; nevertheless, it offers a remarkably high steganographic bandwidth, on average more than 31 kbit/s. However, caution must be taken, as the G.711.0 bitrate is variable and depends on an actual signal being transmitted in the overt channel.

The AMR codec working in 12.2 kbit/s mode proved to be very efficient as a covert codec in TranSteg. This is a low bitrate codec that does not significantly degrade quality: the steganographic cost ranged between 0.36 and 0.46 MOS. However, it is relatively easily detectable using simple analysis when left unmodified. Thus, its header should be modified in order to confuse a potential detector. In contrast, the pairs G.711/GSM06.10 and iLBC/G.723.1 turned out to be quite secure against simple detection methods based on the analysis of the first payload byte. However, further studies are needed to evaluate the resistance against more advanced methods of steganalysis.

7.2 INFORMATION HIDING IN POPULAR P2P SERVICES

P2P services like Skype or BitTorrent are ideal candidate for a hidden data carrier. In the following subsection, we will review two examples of methods for theses P2P services: SkyDe [24] and StegTorrent [25].

7.2.1 Skype

Currently, Skype is one of the most popular IP telephony systems. It is a proprietary P2P telephony service originally introduced in 2003 by creators of the famous P2P file sharing system Kazaa—Niklas Zennström and Janus Friis. Skype is owned by Microsoft and it has been reported that it has about 663 million registered users (September 2011). In January 2013, it was reported to have about 50 million users online simultaneously. It has been also estimated in TeleGeography Report [26] that in 2013, Skype had acquired about 36% of the world's international telephone market. Therefore, because of its popularity and traffic volume, Skype traffic is an ideal candidate for a secret data carrier.

Skype is a hierarchical P2P network with a single centralized element—a login server that is responsible for the authentication of Skype nodes before they access the Skype network. Moreover, the P2P network is formed by two types of nodes [27]: *(i)* Ordinary Nodes (ONs) that can start and receive a call, send instantaneous messages, and transfer Chapters, and *(ii)* specialized nodes called Super Nodes (SNs) that are responsible for helping ONs find and connect to each other within the Skype network.

Skype is based on proprietary protocols and makes extensive use of cryptography, obfuscation, and anti-reverse-engineering procedures. All information about its traffic characteristics, protocols, and behavior comes from numerous measurement studies [28,29].

All traffic in Skype is encrypted using AES (Advanced Encryption Standard)—neither signaling messages nor the packets that carry voice data can be uncovered. Typically, the preferred, first-choice transport protocol of Skype is the User Datagram Protocol (UDP), which as the traffic analysis in [28] showed, is being used in about 70% of all calls. However, if Skype is unable to connect using UDP it falls back to TCP. The steganographic method for Skype presented in this subsection is devoted only to UDP-based Skype calls.

For a TCP-based transport, the entire Skype message is encrypted. In cases of unreliable UDP, at the beginning of each datagram's payload, an unencrypted header is present that is called the Start of Message (SoM). It is unencrypted in order to be able to restore the sequence of packets that was originally transmitted, to detect packet loss, and to quickly distinguish the type of data that is carried inside the message.

Skype estimates the available bandwidth and the packet loss probability, and it dynamically adapts to the detected network conditions by adjusting the codec's bitrate or by introducing higher redundancy in packets [28]. Typically, the resulting packet rate is about 16, 33, or 50 packet/s [27]. It is also important to note that Skype does not utilize any silence suppression mechanism, that is, even if there is no voice activity during the conversation, the packets that carry the silence are still generated and sent. The lack of support for silence suppression is intentional–it helps to obtain better voice quality and maintain UDP bindings for Network Address Translation (NAT) [30].

One can always question why use steganography for Skype that uses encryption to provide confidentiality for every type of user message that is exchanged (text messages, voice and video signals, and files). First of all, hidden communication need not necessarily be conducted in an end-to-end manner, that is, covert data can be piggybacked on third party VoIP calls. Second, Skype calls were commonly believed to be very hard to wiretap, but recently a discussion arose about providing lawful interception services to law enforcement agencies [31]. Additionally, Skype is proprietary and closed software and thus, ultimately cannot be trusted.

Therefore Mazurczyk et al. [24] proposed SkyDe (Skype Hide) method that operates by reusing encrypted packets with silence, by substituting the silence with secret data. The method is feasible because in 2008, Chang et al. [30] observed that in Skype traffic, speech activity is highly correlated with packet sizes, as more information is encoded into a voice packet while a user is speaking. Therefore, a third party can identify silence packets based on the small packet sizes in Skype (Figure 7.4). Experimental results revealed that the packet size and speech volume are highly correlated because they fluctuate in tandem (Figure 7.4).

SkyDe utilizes encrypted Skype voice packets as a hidden data carrier. By taking advantage of the high correlation between speech activity and packet size described earlier, packets without voice signal can be identified and used to carry secret data (by replacing the encrypted silence with secret data bits). As is commonly known, typical VoIP calls contain 35–70% silent periods in each direction [32,33]. Skype does not utilize any silence suppression algorithm; thus, generally, all packets with silence can be utilized for steganographic purposes. However, SkyDe has potentially low impact on voice quality, because it does not affect packets with voice signals, which are more significant from the point of view of conversation.

The proper selection of the level of utilization of packets with silence is necessary, and typically, it will be a trade-off between Skype call quality, desired steganographic bandwidth, and undetectability. Experimental results show that SkyDe offers a high steganographic bandwidth of up to 1.8 kbit/s (for 30% utilization of packets with silence), whilst introducing almost no distortion to the Skype call. Moreover, we prove that under these circumstances the method remains undetectable.

SkyDe can be also used for other IP telephony systems that encrypt their traffic and have a similar relationship between speech activity and packet size. Chang et al. [30] point that this is also a case for UGS (Unsolicited Grant Service).

SkyDe is designed to utilize Skype encrypted voice streams to enable clandestine communication. It must be noted that from the Skype users' perspective detection of SkyDe, as well as steganographic bandwidth, depends on the hidden communication scenario in which it is utilized. This secret data exchange can be realized between (i)

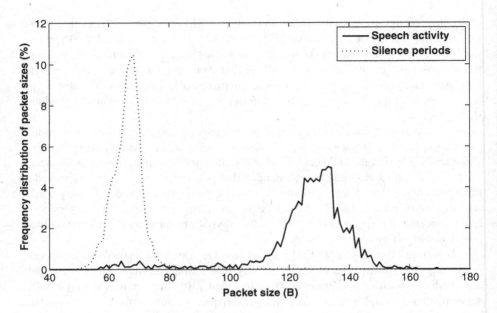

Figure 7.4. The distribution of packets' size during conversation and periods of silence.

two Skype users, that is, they use their own call for steganographic purposes, or *(ii)* a secret data transmitter and receiver that utilize an existing third-party Skype call (in this case, the original caller and callee are not aware of the information hiding procedure). In the first case, users can select the desired steganographic bandwidth because they do not necessarily expect high voice quality. Additionally, they do not care about the elevated overall packet loss level; thus, they are able to transmit about 3 kbit/s. However, in the latter case, the degradation of voice quality or introduced losses cannot be excessive because it could make overt users suspicious. Thus, this could potentially limit the maximum steganographic bandwidth that could be achieved. Therefore, from the point of view of voice quality and packet losses, experimental results revealed that SkyDe will be most undetectable when 30% of packets with silence are utilized (1.8 kbit/s of steganographic bandwidth).

From the network perspective, the byte distribution in the packets' payload is very similar as both the secret messages as well as the original user's voice is encrypted. However, if someone is monitoring Skype traffic, for example, the packet rate and packet sizes, then for SkyDe higher utilization rates of ''silence packets'' could be more easily visible (of course, if the monitoring entity can distinguish between worsening network conditions and hidden communication). Therefore, for this case also, utilization rates that mimic typical Skype calls should be used, that is, up to 30% utilization of packets with silence (almost no difference in the packet rate and sizes from typical Skype calls). Also, for poorer network conditions, less steganographic information could be transferred due to the 70% overall packet losses limit.

In the following two subsections, we will introduce both components of SkyDe, the transmitter and the receiver, separately.

7.2.1.1 SkyDe Transmitter.

First, the secret data is encrypted prior to sending. The cryptographic key utilized for encryption is a shared secret between SkyDe users. Then, a CRC-16 checksum is calculated on the payload and is inserted into the ID field (16 bits) of SoM. Such an approach provides steganogram error detection and facilitates the identification of packets that carry secret data at the receiving end.

Then, those packets that contain silence must be identified. Their size can change while the connection lasts (due to network conditions), so we propose a "sliding time window" algorithm that calculates the reference size of the packets with silence. The algorithm is also responsible for keeping the total packet losses at a safe level, because as mentioned earlier, steganographic utilization of each packet with silence will increase the overall packet loss for the call (the modified packets are deemed lost by Skype). Thus, it is important not to exceed a threshold of about 70% for total packet losses because then the connection will fall back to TCP.

The proposed algorithm works as follows. First, the size of the "sliding time window" w (in seconds) is selected, in which the reference size value r of Skype packets with silence is continuously updated during the call. Every second, the packet with the lowest size is determined and stored. When the window is filled with w packet sizes, the average reference value is calculated based on the three packets with the lowest size. Additionally, a certain deviation in packet size (Δ) $r \pm \Delta$ is acceptable for SkyDe purposes. In the same time window w, it is also verified whether the total packet loss level is 70% or above. If this threshold is reached, SkyDe temporarily ceases to utilize packets with silence until the losses are again at a safe level.

After the packets with silence are identified, their payloads are replaced with encrypted secret data and they are sent to the receiving side. To provide reliability for SkyDe, an additional protocol in a hidden channel might be required. One solution is to use an approach proposed by Hamdaqa and Tahvildari [34] because it can be easily incorporated into SkyDe. It provides a reliability and fault-tolerance mechanism based on a modified (k, n) threshold of a Lagrange Interpolation and the results demonstrated in that paper prove that the complexity of steganalysis is increased. Of course, the "cost" for the extra reliability is a reduced steganographic bandwidth.

7.2.1.2 SkyDe Receiver.

At the receiver side, each packet with silence is recognized by the same means (packet size) as at the transmitter side. These packets are then copied to the buffer and for each of them, a CRC-16 checksum is verified and the secret data is extracted. It is not important to erase or replace secret data embedded into packets because the receiving Skype client involved in a call will treat these packets as losses. However, it must be noted that the modified SkyDe packets will not be discarded while traveling through the network–only the receiving end will be able to identify the corrupted packets.

7.2.2 BitTorrent

BitTorrent ([35,36]), a file-sharing system originally released in July 2001, is currently the most popular P2P (Peer-to-Peer) file-sharing system worldwide. Studies show that the number of users exceeded 100 million in 2011, and that BitTorrent traffic accounts for about 94% of all P2P traffic. The success of BitTorrent primarily comes from two factors: its efficiency and openness. BitTorrent is significantly more efficient than classical client/server-based architectures. It allows peers sharing the same resource to form a P2P network, and then it focuses on fast and efficient replication to distribute the resource. Additionally, BitTorrent software is free to download and many clients are open source. This leads to the easy deployment of new applications and technologies, therefore stimulating further improvements.

BitTorrent is distinguished from other similar file-transfer applications in that instead of downloading a resource (one or more files) from a single source (e.g., a central server), users download fragmented files from other users at the same time. As a result, the file-transfer time is considerably decreased because the group of users that share the same resource (or part of it) may consist of several to thousands of hosts. Such a group of users interested in the same resource (known as *peers*) combine together with a central component (known as a *tracker*) in BitTorrent. This combination of peers and trackers is called a *swarm*. Trackers are responsible for controlling the resource transfer between the peers. Peers that hold onto a particular resource or part of a resource are required to share the resource and to perform the transfer. We can distinguish two types of BitTorrent peers based on the stage at which they are involved in downloading or sharing a given resource: *(i)* seeds are peers that possess the complete resource and are only sharing it and *(ii)* leechers are peers that do not possess the complete resource and are still downloading the missing parts. All peers share the fragments they have already downloaded. When a leecher obtains all the remaining fragments of the resource it automatically becomes a seed. It is also worth noting that because in BitTorrent a resource is divided into many fragments, a single peer is able to download many fragments simultaneously and it does not need the whole resource to share it.

In the BitTorrent specification, two main protocols are described that regulate data transfer: *peer-tracker* and *peer-peer*. The connection between peer and tracker can be established with the use of HTTP (Hypertext Transfer Protocol) or through UDP-based requests. The tracker is used mostly to initiate the connection with a swarm. After the connection is established, popular BitTorrent extensions like PEX (Peer Exchange) [36] or DHT (Distributed Hash Table) [37] are used. These extensions enable communication between peers without using a tracker as part of the swarm.

According to the BitTorrent specification, the peer-peer data exchange should be conducted using a proprietary application layer protocol on top of TCP. It is a stateful protocol that is used to establish connections similar to the TCP handshake mechanism. However, besides the specified protocol over TCP, since 2009 there is also a UDP-based protocol called μTP (μTorrent Transport Protocol) [38]. μTP is not part of the original BitTorrent specification (however, it was created by BitTorrent Inc.) but as the results in [39] indicated, it is currently the most popular protocol.

The main aim of the μTP protocol is to efficiently manage usage of the available bandwidth during file transfers, while limiting the impact of file transfers on the on-going transmissions (especially non-BitTorrent related ones). The μTP protocol is capable of automatically reducing the rate at which BitTorrent packets are transmitted between peers in case there is an interference with other applications running on the same host. μTP uses a congestion control algorithm, which is a modified version of LEDBAT (Low Extra Delay Background Transport) [40], based on one-way delay measurements. The μTP protocol is used by default in many popular BitTorrent clients like μTorrent beginning with version 2.0, BitTorrent, Vuze, or Transmission.

Typically, the BitTorrent client establishes many connections with other peers during the whole transmission process—typically about 70 [39]. Due to BitTorrent's popularity, high but variable packet rates, and the fact that a single resource is downloaded by a number of clients simultaneously (one-to-many transmission), it is very suitable as a carrier for hidden data exchange. The one-to-many transmission feature is used by the steganographic method StegTorrent [25] to enable clandestine communication. StegTorrent relies on the modification of the order of the data packets in the peer-peer data exchange protocol. StegTorrent takes advantage of the fact that the μTP header provides a means for packet numbering and retrieval of their original sequence and manipulate this information to influence the order in which packets will be processed at the receiving end. This allows a high steganographic bandwidth (hundreds of bits per second), while remaining undetectable with no requirements on sender–receiver synchronization (which is often necessary for timing methods). The StegTorrent steganographic cost is negligible, since it only introduces a small delay in resource downloading by resorting the packets in the predefined order. This is completely transparent to the BitTorrent client.

The clandestine communication scenario we consider for the proposed method is illustrated in Figure 7.5. We assume that both the covert sender and receiver are in control of a certain number of BitTorrent clients and, as mentioned above, their IP addresses

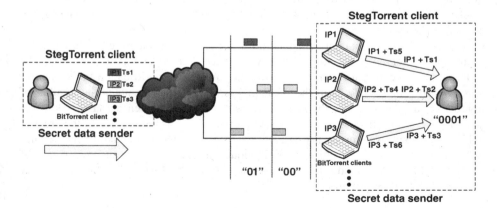

Figure 7.5. StegTorrent hidden data exchange scenario (Ts*X* denotes a timestamp from the corresponding μTP header's field. (Reproduced from [25] with permission of IEEE.)

are known to each other. In Figure 7.5, for the sake of clarity, only a single direction steganographic transmission is presented, but of course, bidirectional communication is possible and the other direction is analogous. No knowledge of the network's topology is necessary. The secret sender uses the modified BitTorrent client—StegTorrent client—to share a resource that is downloaded by the covert receiver using a second StegTorrent client that consists of a group of controlled BitTorrent clients.

For the sake of the proposed method's description and analysis, we define the term *data package* as a set of IP addresses that is sent within the IP packets in a predetermined order and the term data package size as the total number of elements in this set. For example, let us assume that the data package size is 2. In this case, two packets with two different destination IP addresses (e.g., IP1 and IP2) are used to send bits of hidden data. The order of packets is modified to encode the covert bits, for example, the covert sender encodes a binary "0" as a packet sent to IP1 followed by a packet sent to IP2, and a binary "1" is encoded as the opposite order. We assume that the data package and its size are a shared secret between transmitting and receiving StegTorrent clients. It must be noted that this method's performance depends on the size of the data package that in turn relies on the number of available receiving IP addresses (receiving BitTorrent clients under control).

Experimental results showed that StegTorrent is feasible and it provides a steganographic bandwidth of about 270 bit/s for the most realistic tested scenario while being hard to detect.

In the following two subsections, we will introduce two vital parts of StegTorrent covert communication, that is, the transmitter and the receiver.

7.2.2.1 StegTorrent Transmitter.

At the transmitting side the secret data bits are encoded in the order in which the data packets are sent to the particular set of receiving clients. To encode secret data bits a coding scheme like a Lehmer code [41] can be utilized (the encoding is known *a priori* to the secret sender and receiver). Then for n different numbers of IP addresses, we are able to encode $\lfloor log_2(n!) \rfloor$ bits per data package. For example, for four packets with different IP addresses in a data package we are able to send 4 bits.

However, due to poor network conditions, the order of the packet could be changed. This could potentially lead to problems with successful steganogram extraction. The solution is to incorporate into StegTorrent the intentional modification of the *Timestamp_microseconds* field from the μTP protocol. This field contains the number of microseconds elapsed since the last full second of the time when the last packet was sent. Hence, values from these fields allow the unambiguous recognition of the order in which the packets were originally transmitted. By intentionally modulating these values while generating packets to enforce certain sequences, the sender can ensure that, even if the order of the transmitted packets is disrupted, the receiver would be able to correctly extract bits of the steganogram.

7.2.2.2 StegTorrent Receiver.

The secret receiver gathers all the information from the BitTorrent clients under the control of StegTorrent during the resource download session. Then, it orders the packets' IP addresses based on their *Timestamp_microseconds* values and begins the extraction of secret data bits.

If some packets are lost then the secret receiver will wait for retransmission, which is ensured by the μTP protocol. The retransmitted packets will have the same *Timestamp_microseconds* values as the original ones. After the lost packets have been retransmitted, the secret receiver can extract the secret data.

7.3 INFORMATION HIDING IN MODERN MOBILE DEVICES[1]

As mentioned in Chapter 1, in recent years, the rapid technological advances of software and hardware led to mobile devices offering capabilities previously achievable only for desktop computers or laptops. The increasing convergence of network services, computing/storage functionalities, and sophisticated Graphical User Interfaces (GUIs) culminated into new devices called *smartphones*, which are quickly becoming the first choice to access the Internet, and for entertainment. Moreover, the Bring Your Own Device (BYOD) paradigm makes them important tools in the daily working routine.

Globally, it is estimated that there are about 5 billions of mobile phones worldwide, of which ~1.08 billion are smartphones. [2] It must be emphasized that a relevant pulse to this increase is caused by worldwide smartphone shipments that are expected to surpass 1 billion units for the first time in a single year. This popularity is mainly driven by a multi-functional flavor combining many features, such as a high-resolution camera, different air interfaces (e.g., 3G, Bluetooth, IEEE 802.11, etc.), and Global Positioning System (GPS) into a unique tool. Another key reason of this huge success is the advancement of cellular connectivity, allowing users to interact with high-volume or delay-sensitive services while moving, for example, through the Universal Mobile Telecommunications System (UMTS), or the Long Term Evolution (LTE).

In fact the possibilities for data hiding are dramatically increased in smartphones for the following reasons:

- the multimedia capabilities enable to create and use a wide variety of carriers, such as audio, video, pictures, or Quick-Response (QR) Codes;
- the availability of a full-featured TCP/IP stack, as well as the possibility of interacting with desktop-class services lead to a complete reutilization of all the network methods already available for standard computing devices or appliances; and
- the richness of the adopted OS permits to develop sophisticated applications, thus making covert channels based, for example, on VoIP and P2P exploitable.

---

[1] The following section is based on the publication ''Steganography in Modern Smartphones and Mitigation Techniques'' by W. Mazurczyk and L. Caviglione [42]

[2] http://www.go-gulf.com/blog/smartphone/

The architecture of the current mobile OSes forces steganographers to search for ways not only to exfiltrate data using network but rather to acquire the data to be exfiltrated and then gain access to the network [42]. This is somewhat complicated due to enforced security policies used in mobile OSes, for example, many manufacturers enforce users to install software only from verified sources (e.g., iOS which only permits software provided by the Apple AppStore) that deeply inspect the software before the publishing stage, and thus detect malicious threats early.

Therefore, network steganography in smartphones is usually jointly used with a local covert channel. While the former can be implemented by utilizing known methods from desktops, the latter is the actual technological enabler for data exfiltration. In fact, this is a typical scenario considered for smartphones but of course covert channels can be utilized on mobile phones for other purposes too. The typical scenario for local covert channels considers two processes, P1 and P2, each one running in a sandbox S1 and S2, respectively. Specifically, S1 and S2 have been configured to prevent data leaking. For instance, if P1 can access the address book but not the network, S1 will impede to use network interfaces. Let us assume then P1 is a malware wanting to exfiltrate the address book, while P2 is companion application (often named a colluding application) looking innocuous. This security mechanism is typically bypassed as follows: P1 leaks data to P2 via a covert channel, and P2, after receiving the data, will use network steganography to contact a remote facility in a stealthy manner.

The first work implementing such a mechanism is a malware called SoundComber [43], which is able to capture personal user data, like the digits entered on the keypad of the device during a phone call. It can use different methods to enable the two colluding processes to communicate by embedding secrets in well-defined patterns of toggling the vibration, hiding data within changes in the volume level of the ringtone, or locking/unlocking the screen. Other works that investigated above-mentioned scenario for sensitive user data exfiltration include papers by Marforio et al. [44] and Lalande and Wendzel [45].

When secret data are ready to be exfiltrated then a network steganography method is needed. One of the first technique that allowed to form a network covert channel for smartphones was proposed by Schlegel et al. [43], and it is based on using a web browser to send requests to web servers, where the URLs are of the form http://receiver?number=N, where N is substituted with the secret data string to be exfiltrated to the target website.

Gasior and Yang [46] proposed two ideas to implement a network covert channel. In the first method, an application transmits live video from the camera of an Android device to a remote server. Then, the steganographic method creates a covert channel by altering the delays between the video frames sent to the server (which is in fact the implementation of a known approach in the mobile context—see Chapter 3.2). The experimental evaluation was performed in [47] and the resulting steganographic bandwidth is ~8 bit/s. The second covert channel uses a tool that displays an advertisement banner at the bottom of a running application. The content of the banner is selected from N different advertisements available, and then requested from the remote server. The application leaks sensitive information, for example, the contact list, by encoding the

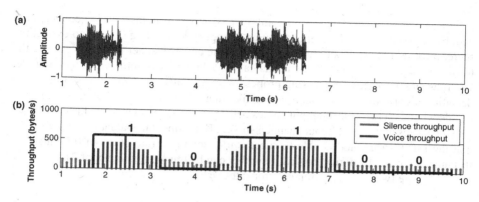

Figure 7.6. iStegSiri's crafted voice stream (a) results in corresponding classes of traffic (light gray–voice, dark gray–silence), which enables successful detection of secret bits at the receiving side (b). (Reproduced from [48] with permission of IEEE.)

sensitive data as series of requested adds, where each add is associated with a binary value.

Caviglione and Mazurczyk were the first to propose a steganographic method for iOS-based devices called iStegSiri [48]. It exploits the built-in Siri service that allows to interact with an iPhone/iPad using voice commands. Siri was originally developed as a standalone application in 2010 and offered as a native service from iOS5 in 2011. In Siri, to offload the smartphone, the translation of voice inputs to text is performed remotely in a server farm operated by Apple. The user-generated voice samples are sent from a mobile device to a remote facility and then a response containing the recognized text, a similarity score and a time stamp are returned.

iStegSiri utilizes the architecture and functionality of Siri and the discovered correlation between generated voice patterns and resulting traffic characteristics [49]. An attacker could produce ad-hoc voice patterns to manipulate the throughput and encode a secret into the resulting traffic that is illustrated in Figure 7.6. To achieve this, she first encodes a secret message into an audio sequence using proper modulation of voice and silence periods Figure 7.6(a). Then the generated sound pattern is provided as the input to Siri (using, for example, the internal microphone). As a result, the device will send traffic to the remote server that does the audio to text conversion. Unfortunately, algorithms used for synchronization, reduction of latencies and packetization delays, prevent forging the throughput of the whole flow, even with a minimal degree of accuracy. To overcome this drawback, it is possible to split the overall traffic into different components by using a set of ranges for the Protocol Data Units (PDUs) produced by Siri. Specifically, PDUs in the range of 800–900 bytes were effective in representing talk periods, while PDUs in the range 100–700 bytes, that is, silence periods. With such a partition, it is possible to arbitrarily encode binary bit values within the traffic throughput, that is, by alternating talk/silence periods as to increase/decrease the number of PDUs belonging to each defined range (see Figure 7.6(b)). In trials, the shortest working values were 1 s and 2 s, for voice and silence, respectively [48].

Next, the recipient of the secret communication passively inspects the conversation and, by observing a specific set of features, it applies a decoding scheme to extract the secret information. This is only possible when the covert listener is able to capture the traffic and decode the secret. The former can be achieved in several ways, for example, via transparent proxies, or by using probes dumping traffic for offline processing. The decoding algorithm implements a voting-like method using two decision windows to decide whether a run of throughput values belongs to voice or silence, that is, 1 or 0 (see also the basics of the throughput channels in 3.2). Experimental evaluation demonstrates that the iStegSiri method is able to successfully send at a rate of about 0.5 bit/s. For instance, a typical 16-digit credit card number can be transmitted in about 2 min.

There are two main requirements for iStegSiri:

- An attacker must be able to access the internals of the service. For the case of iOS it means that only jailbroken devices can be utilized through a library called libActivator or by directly accessing the private APIs provided by Apple. The "audio track" used to encode the secret can be produced by the malware at runtime, that is, by replicating a single sample via software, thus without requiring to inflate the size of the executable. The audio data can be directly routed from the malware to the codec, without requiring a playback audible for the user [3].
- An attacker must be able to access the steganographically modified Siri traffic while it travels to Apple's server facilities (several ways to achieve this were mentioned above).

7.4 INFORMATION HIDING IN NEW NETWORK PROTOCOLS

Network steganography methods have been proposed not only for existing, well-known and mature network protocols, but also for those that are expected to rule communication networks in the near future. Preventing creation of covert channels during its design phase or at an early stage of its deployment is considered as one type of possible countermeasures against information hiding (see Chapter 8 for details).

An example of such rigorous analysis of the IPv6 protocol was conducted in 2006 by Lucena et al. [50] where 22 potential information hiding methods were identified. However, the study considered only storage methods and the described techniques were classified according to the type of header. To defeat the identified channels, three types of active wardens were defined stateless, stateful, and network-aware, which differ in complexity and ability to block some types of covert channels (see Chapter 8 for details on various types of wardens). Later, Mazurczyk and Szczypiorski [51] proposed two storage steganographic methods for Path MTU Discovery (PMTUD) and Packetization

---

[3] In experiments a microphone was used for iStegSiri evaluation.

Layer Path MTU Discovery (PLPMTUD) mechanisms that are utilized in IPv6 to deal with oversized packets.

A similar comprehensive study was performed by Frączek et al. [52] for Stream Control Transmission Protocol (SCTP). SCTP is a candidate for a new transport layer protocol that may replace the TCP (Transmission Control Protocol) and the UDP (User Datagram Protocol) protocols in future IP networks. Currently, the SCTP is implemented in, or can be added to, many popular operating systems, for example, Windows, BSD, Linux, or Oracle Solaris. SCTP provides some of the same service features of TCP and UDP protocols, ensuring reliable, in-sequence transport of messages with congestion control. However, its main advantages include capabilities of multistreaming (i.e., an ability to use one or more streams to transfer data that are unidirectional logical channels between SCTP endpoints) and multihoming (i.e., an ability of a host to be visible in the network through more than one IP address that increases reliability of data transfer). A rigorous analysis presented in [52] revealed steganographic vulnerabilities in these new features and characteristics of SCTP. In particular, 19 different storage and timing covert channels were identified. This study can be treated as a guide when developing countermeasures for SCTP steganography. In fact, it also points some detection/prevention solutions as many of the proposed hiding methods can be evaded by simply modifying the current standard.

7.5 INFORMATION HIDING CONCEPTS FOR WIRELESS NETWORKS

The family of IEEE 802.11 standards has become the most popular method of the wireless Internet access as well as the way of organizing Small Office Home Office (SOHO) networks. Laptops, smartphones, printers, tablets, home theaters, TVs, radio sets, and even WiFi-enabled bulbs can communicate without wired connection, but usually still need power cables (it cannot be fully wireless itself).

Unfortunately, the popularity of WiFi networks, in addition to the regular benefits like easy configuration and significant bandwidth, also brings some risks. One of them is the simplicity of radio channel eavesdropping as a typical feature of shared medium networks: in WiFi networks the members have the possibility of "hearing" all the data frames exchanged in the air and the only limit is to be in a range. For network steganography, this feature seems to be attractive—broadcast communication (one to all) opens the mind of eager steganographers to the new scenarios of hidden communication without boundaries. At a first glance it looks very promising, but for WiFi networks "a range is the limit" and typically is not larger than 100 meters (330 feets) from a source station. However, it must be noted that techniques for WiFi steganography can be useful for constructing hidden channels as well for "long-distance" wireless technologies like Long-Term Evolution (LTE) or Worldwide Interoperability for Microwave Access (WiMax). Below we review the most well-known information hiding techniques for wireless networks based on what characteristic feature they exploit. We also highlight existing efforts to implement these steganographic methods.

7.5.1 Intentionally Corrupted Checksums

Szczypiorski proposed HICCUPS (*HIdden Communication system for CorrUPted networkS*) [53], a first steganographic system for WiFi networks [54]. The main innovation of the system is the usage of frames with intentionally corrupted checksums to establish covert communication. When a wireless client detects an error in a broadcasted frame, it simply drops that corrupted frame. The error-checking mechanism is based on the Frame Check Sequence (FCS), a kind of "signature" against which the integrity of the packets can be confirmed. When the receiver gets a packet, it checks for errors using that packet's checksum. In HICCUPS, frames with intentionally corrupted checksums form a steganographic channel.

To detect HICCUPS, one need some way of observing the number of frames with incorrect checksums. If the number of those frames is statistically anomalous, then one might suspect the transmission of hidden information. Another way of detecting HICCUPS is to analyze the content of those dropped—and therefore retransmitted—frames in order to detect the differences between the dropped and retransmitted frames. Szczypiorski [55] proved that HICCUPS bandwidth equals 1.27 Mb/s for an IEEE 802.11g 54 Mb/s network with 10 stations and grabbing 5% of traffic for steganography. HICCUPS is not easy to implement in hardware, because FCSes are calculated on board of network cards. It is possible, but not proven, to build a HICCUPS-enabled card on the top of a field-programmable gate array (FPGA) card with capabilities of the software-defined radio.

Najafizadeh et al. [56] presented a simulation of HICCUPS based on an initial work of Odor et al. in [57].

7.5.2 Padding at Physical Layer

The WiPad (Wireless Padding) [58] by Szczypiorski et al. is dedicated to the 802.11a/g standards and utilizes that some WiFi networks, in particular newest ones, rely on the data-encoding technique known as orthogonal frequency-division multiplexing (OFDM). In the OFDM scheme, several small-bandwidth carriers of different frequencies are used to send data over the air. These narrowband carriers are more resilient to atmospheric degradation than a single wideband wave, allowing data to pass to receivers with higher fidelity. OFDM carefully selects carriers and divides the bits up into groups of set length, known as symbols, to minimize interference. In real-world scenarios, a frame rarely divides perfectly into a collection of symbols; there will usually be some symbols left with too few bits. So, OFDM transmitters add extra throwaway bits, called "bit padding", to these symbols until they conform to the standard size. Because this "bit padding" is meaningless for upper layers, it can be replaced with secret data without compromising the original data transmission.

The investigation of the IEEE 802.11 frame reveals that two other fields are also liable to padding: SERVICE and TAIL. The lengths of SERVICE and TAIL are constant (16 and 6 bits, respectively), while a Physical layer Service Data Unit (PSDU) is a medium access control (MAC) frame and its length depends on user data, ciphers and network operation mode (ad hoc versus infrastructure). The padding is present in all frames, therefore some frames that are more frequently exchanged, like ACKs may become an interesting target for covert communication. From the steganographic bandwidth perspective it means that for an 802.11g network with 54 Mb/s rate (64-QAM, $\frac{3}{4}$ code rate) the symbol consists of 216 bits. Having 216 octets frames means that the resulting steganographic channel has a capacity 1.1 Mb/s and additionally 0.44 Mb/s could be utilized from ACK (acknowledgment) frames. This sums to 1.54 Mb/s in total.

The OFDM padding study [58] led to similar efforts on LTE [59] and WiMax [60] by Grabska et al.

7.5.3 Controlling the Intervals Between OFDM Symbols

The work on WiPad was further continued and extended by Grabski et al. [61] to explore other features of OFDM symbols and to apply information hiding mechanisms for wireless high speed networks like 802.11n. In the radio environment of wireless networks, the multipath propagation results that the receiver captures not only the signal propagated directly from the source, but also its delayed copies. Consequently, transmitted OFDM symbols could be affected by reflected signals from obstacles, called intersymbol interferences. In order to reduce these interferences, a special guard interval is inserted between each pair of OFDM symbols. IEEE 802.11 standards implement filling that protection gap with the cyclic prefix. In order to hide information, the system proposed by Grabski et al. changes the cyclic prefixes of the chosen OFDM symbols and turns them into fragments of the secret message. The achievable steganographic bandwidth for hidden channel ranges from 3.25 to 19.5 Mb/s.

7.5.4 Proposals Specific to 802.11 Networks

Some hiding techniques exploited issues specific to 802.11 networks:

- CSMA/CA (Carrier Sense Multiple Access with Collision Avoidance) [62],
- Controlling the rate of transmission [63],
- Beacon mechanism [64].

Holloway [62] proposed Covert DCF, a timing channel based on the CSMA/CA mechanism used in 802.11 to avoid collisions. The main idea of the proposed scheme was a controling of random backoff—an internal mechanism for counting a time for the next transmission. The author was able to achieve a steganographic bandwidth around 2 kb/s.

Calhoun et al. [63] proposed a side channel that uses the 802.11 MAC rate switching protocol to hide communication between an access point and a station. They explored two applications for this channel: covert authentication and covert WiFi botnets. The covert authentication proposed by Calhoun et al. [63] was further developed by Sawicki et al. [64] to authenticate access points by using beacon frames and timestamp fields.

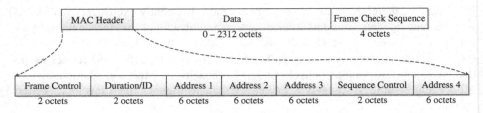

Figure 7.7. The structure of the MAC frame. (Reproduced from [67] with permission of IEEE.)

Protocol version	Type	Subtype	To DS	From DS	More fragments	Retry	Power management	More data	WEP	Order
2 bits	2 bits	4 bits	1 bit	1 bit	1 bit	1 bit	1 bit	1 bit	1 bit	1 bit

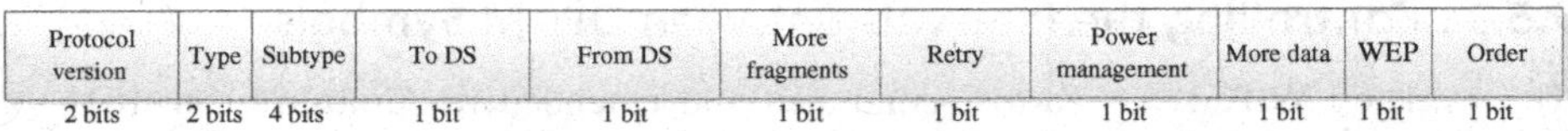

Figure 7.8. Frame Control field. (Reproduced from [67] with permission of IEEE.)

7.5.5 Implementations: Rather Modification of a Header

Some of the steganographic channels were implemented in software [54,65,66]. Almost all of them are based on the modification of a header (Figure 7.7), which is the easiest way of controlling this part of the frame. Grabski et al. [67] also proposed a tool for steganalysis.

Krätzer et al. [54] proposed two steganographic methods based on packets duplication/modification. The first method is based on header modification, especially on the "Retry" and "More Data" bits and on the "Duration/ID" field (Figure 7.8). The first is used for synchronization, and the second for transmitting the hidden data. The second technique is based on the affection on time dependencies between transmitted frames. The work was slightly revised by the authors in [68].

Frikha et al. [65] created a steganographic system based on the header modification at the MAC sublayer. The authors proposed to use independently two fields of a MAC header: Sequence Control (SC) and Initial Vector (IV) for WEP (Wired Equivalent Privacy)-protected frames. The 16 bits SC field is divided into two subfields: sequence control (incremented by every new frame) and fragment control (incremented by every new fragment if the frame is fragmented); the first 12 bits of the field are used for a hidden channel. The IV field is 24 bits long, and should be random. It is worth noting that WEP encryption is currently not commonly used as it is deemed insecure. Goncalves et al. [66] extended the work of Frikha et al. and proposed to use two bits in the protocol version field of the Frame Control Field in a MAC header to carry hidden information. Additionally, the authors introduced the use of forward error correction and bit interleaving mechanisms to improve the performance of the given hidden channel.

Finally, Grabski et al. [67] implemented a WiFi steganalyser as a tool to monitor the network traffic passively in order to detect hidden communication. The system recognized five scenarios based on [54] and [65] including these based on the "Duration/ID" field, as well as on "Retry" and "More Data".

7.6 MULTIPLAYER GAMES AND VIRTUAL WORLDS[4]

Over the last two decades network multiplayer games have become popular in the Internet. These days many games exist where different human players control their virtual alter egos (called *avatars*) and interact with other players in virtual 2D or 3D worlds. Network steganography can be used to covertly exchange information between two or more players in a virtual world.

Multiplayer games or other virtual worlds typically offer several internal communication channels, such as text chat or voice communication. However, these overt communication channels can be easily monitored, for example, text chat is usually logged and filtered at servers. Network steganography allows hiding the communication from server operators and unwitting players. Even players whose clients are endpoints of the covert channel may remain unaware of the covert information flow.

The steganographic technique we describe in this section was initially developed for First Person Shooter (FPS) games and named FPS Covert Channel (FPSCC) [69]. However, the principles behind the approach are fairly general and could be applied to other games or virtual worlds where players control avatars.

FPSCC hides covert information in very small, additional movements of avatars. Movements intended by a human player are slightly varied to encode covert bits. As long as the variations are small, they will have no visible effect on the avatar's movements as perceived by other human players inside the virtual game world. Avatar movement is often not logged at servers, hence the information transfer is covert. Since movement of avatars is an intrinsic function of multiplayer games and virtual worlds, it is impossible to remove the channel other than by preventing the use of these applications or blocking their overt traffic.

7.6.1 First Person Shooter Games

FPS games are a popular genre of multiplayer games and their network traffic is not suspicious, although it may not be present everywhere, for example, game traffic may be blocked by a company firewall. FPS games, such as Quake III Arena, or Counter-Strike Source, are based on a client-server architecture. The game publishers often release their server implementations for free and rely on Internet Service Providers (ISPs), dedicated game hosting companies and individuals to host FPS servers. FPS servers host from under ten to 30+ players, and for popular games there may be tens of thousands of game servers active on the Internet at any given time [71].

Hence, a covert sender and a covert receiver have a wide variety of game servers through which to establish innocent-looking overt traffic flows. Since they use the game server as an intermediary, they do not need to exchange network traffic directly. Also, since information from one player is transmitted to several other players via the server, the channel is actually a broadcast channel. Detection of the covert sender does not

---

[4] This section is based on publications by Zander et al. [69,70].

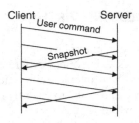

Figure 7.9. Client server message exchange in First Person Shooter games. (Reproduced from [70] with permission of Springer.)

directly reveal the identities of the covert receiver(s), who could be any of the other players online at the same time.

7.6.2 In-game Client-Server Message Exchange

To understand how hidden communication in FPS games works, one first needs to understand the basic concepts behind FPS network protocols.[5] FPS games are based on a client-server architecture and rely on UDP/IP packets to carry information. We focus on the network traffic that occurs during a game (as only this is utilized by FPSCC), and ignore traffic associated with server-discovery and initial client connection.

During a game, each client receives client-specific game world state updates from the server (which are called *snapshots*) as shown in Figure 7.9. A snapshot contains the server's authoritative belief about the state of the client's player (position, view angles and events) and the state of all other entities potentially visible to the client's player. Entities can be other human players' avatars, computer-controlled characters, or objects in the world. Usually, entity state updates are not sent for entities that the client's player cannot see. However, not all potentially visible entities are actually visible on the player's screen. Server messages can also contain commands to be executed by the client, such as printing in-game messages.

Based on the state of the virtual world, players will perform actions that are sent to the server in *user commands*. User commands contain movement information based on keyboard and mouse input. Figure 7.10 illustrates the player movements that may be sent in each command. Movement occurs along three axes (left/right, forward/backward, up/down) and change of view angle occurs along two axes (yaw and pitch).[6] User commands also contain movement-unrelated mouse button and keyboard state, for example, the weapon selection of the player.

Message sequence numbers are used in both directions to detect packet loss and reordering. Explicit retransmissions only happen for commands that need to be reliably transmitted, for example, a disconnect message sent by a client to the server. Most user

---

[5] Our explanation is based on Quake III Arena, but many modern FPS games have similar protocols.

[6] In FPS games usually players cannot modify roll.

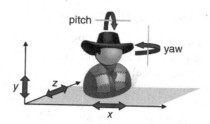

Figure 7.10. Player character movement in FPS games. (Reproduced from [70] with permission of Springer.)

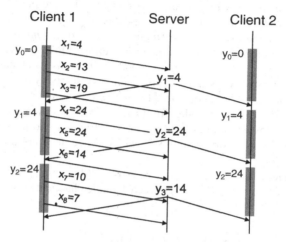

Figure 7.11. An example of user command input values and server snapshot values (Reproduced from [70] with permission of Springer.)

commands, such as movement commands and entity state updates in server snapshots, are continuously updated and never retransmitted. However, there is redundancy to compensate for packet loss. Each UDP/IP packet sent by a client contains the current user command as well as previous user commands. User commands sent by a client are timestamped, as are player and entity state updates sent by a server. Therefore, every update of a player's state sent by a server can be unambiguously linked to a corresponding user command sent previously by a client.

Figure 7.11 illustrates the relationship between player movement information sent to the server, and the same information received by other clients. Let x_i be client 1's player input for their character's position along an axis (or the view angle along an axis) in user command i, and let y_j be the position or view angle of client 1's character sent by the server to both clients in snapshot j. Since user commands usually arrive more frequently than snapshots are emitted, each y_j is computed based on the most recently received x_i. Client 2 displays client 1's player on screen based on y_j, until it receives y_{j+1} (this period is indicated by the boxes).

7.6.3 Encoding and Decoding the Covert Data

FPSCC creates a covert channel between two FPS game clients, where one client acts as a sender (Alice) and the other acts as a receiver (Bob). The covert sender and the covert receiver may be built into actual clients, or they may manipulate game traffic of unwitting players. FPSCC aims to stay unnoticed by the players controlling the game clients, or by a warden (Wendy).

FPSCC only modulates a player's view angle updates for pitch and yaw, since the view angles mostly depend on a player's input only, whereas position information may be perturbed by various other "forces" acting on avatars inside the game's virtual world, making it very hard to predict y_j from x_i. Alice encodes covert information by modulating x_i from client 1 with visually imperceptible fluctuations of an avatar's view angles. Bob decodes the covert data from y_j updates arriving in consecutive snapshots.[7]

Because x_i conveys relative view angles, FPSCC uses *changes* in view angles to encode covert information. To minimize detection, FPSCC only encodes covert information when players are moving their views. The covert channel pauses when players stop changing their view. The covert channel is effectively masked since FPSCC-induced changes are very small compared to the player's input.

Since with FPSCC Bob decodes from all angle changes, Alice must encode in all angle changes. However, a micro-protocol with framing and padding bits between frames can be used to support scenarios in which Alice does not continuously sends data.

Let the change in user input be $\Delta_i = x_i - x_{i-1}$. Alice encodes N covert bits with an integer value of b ($0 \leq b \leq 2^N - 1$) into each angle change so that

$$b = \left| \tilde{y}_j - \tilde{y}_{j-1} \right| \bmod 2^N, \tag{7.2}$$

where $\tilde{y}_j$ and $\tilde{y}_{j-1}$ are the angle values manipulated by Alice. However, Alice can only indirectly modify y_j by modifying the user input x_i.

As explained previously, the game server computes y_j from the most recently arrived x_i. The asynchronous message transmissions, unpredictable client message rate and variable network delay make it impossible for Alice to predict the x_i that will be used by the game server to compute y_j. Therefore, Alice has to encode the same covert bits in all x_i sent between the arrival of $\tilde{y}_{j-1}$ and $\tilde{y}_j$.

When Alice detects an angle change ($\Delta_i \neq 0$), she starts encoding the next covert bits to be sent, b_n, in the current and following user commands. Each time a snapshot is received from the server, Alice checks whether the angle value has changed ($\tilde{y}_j \neq \tilde{y}_{j-1}$). If it has not changed, Alice continues sending b_n. Otherwise, Alice assumes b_n has been successfully transmitted, and she updates the previous angle value $\tilde{y}_{j-1}$. The next user angle change will allow Alice to start sending bits b_{n+1} and so on.

---

[7] FPSCC is not limited to unidirectional communication; Bob could send to Alice at the same time.

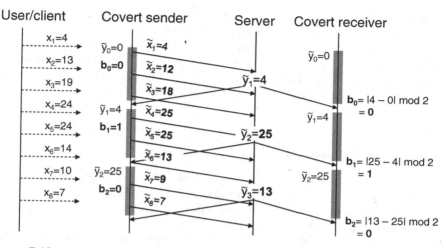

Figure 7.12. Example of covert channel encoding. (Reproduced from [70] with permission of IEEE.)

Covert bits are encoded by modifying user inputs as follows. We define the user input modified by the covert sender as

$$\tilde{x}_i = x_i + \delta_i. \tag{7.3}$$

If $\Delta_i \neq 0$ Alice encodes b by selecting δ_i such that

$$b = \left| \tilde{x}_i - \tilde{y}_{j-1} \right| \bmod 2^N. \tag{7.4}$$

From equation 7.3 and equation 7.4 follows:[8]

$$\delta_i = \begin{cases} b - \left(x_i - \tilde{y}_{j-1} \right) \bmod 2^N & x_i - \tilde{y}_{j-1} \geq 0 \\ -b - \left(x_i - \tilde{y}_{j-1} \right) \bmod 2^N & x_i - \tilde{y}_{j-1} < 0 \end{cases}. \tag{7.5}$$

The next snapshot value $\tilde{y}_j$ will be based on one of the $\tilde{x}_i$ arrived at the server between snapshot $j - 1$ and j, and possible noise on the channel n_j. Bob decodes the covert bit(s) similar to equation 7.4:

$$\hat{b} = \left| \tilde{y}_j - \tilde{y}_{j-1} + n_j \right| \bmod 2^N. \tag{7.6}$$

Figure 7.12 illustrates the encoding of covert bits in one view angle, with $y_0 = 0$ and the same user input as in Figure 7.11. One bit of covert information is encoded per angle change, which means an even change signals a zero-bit and an odd change signals a one-bit. The angles modified by Alice are shown in bold. The boxes indicate the time periods in which a covert bit is transmitted.

---

[8] Also, Alice must avoid completely negating a player's angle change as described in [69].

Figure 7.12 assumes that the round trip time (RTT) between the client and server, plus the time between two client messages, is smaller than the time between two server updates. If this is not the case, Alice never knows the actual value of $\tilde{y}_{j-1}$ and cannot compute the correct δ_i. FPSCC can function over larger RTTs, but the covert bit rate must be reduced. If u is the time between game server updates, FPSCC can send b_n in m server updates intervals, where $m \geq 2$ and $m \cdot u \geq$ RTT.

7.6.4 Channel Noise

FPSCC suffers from three types of bit errors: *substitutions* (bits changed), *deletions* (bits completely lost), and *insertions* (bits inserted).

To reduce the number of objects rendered, reduce network traffic, and mitigate cheating (see Chapter 7 in [71]), snapshots only send the state of potentially visible entities to a player. Potential visibility is determined from visibility information in the map data and the actual positions of players and entities on the map. Potential visibility may be asymmetric, for example, Bob may not receive Alice's state at a time when Alice receives Bob's state, which can cause bit insertion and deletion errors [69].

Players respawn at various map locations after they die or when they enter teleportation devices. Respawning forces a change in the player's view angles, which causes substitution errors, if the dead or teleported player respawns within the visible range of the other player. Rotating platforms, on which avatars can stand on, also introduce view angle changes unrelated to the player's input, which possibly introduces bit substitutions. Further errors can be introduced by server-side angle clamping [69].

The protocol sequence numbers can detect reordered user commands or snapshots due to lost or reordered UDP/IP packets. However, late user commands or snapshots are simply ignored by clients and servers, and so reordered packets may be effectively lost.

User commands are redundant, as several are sent between two snapshots. If no user commands reach the server between snapshots, Alice can not send any covert bits. However, if some user commands arrive, it is crucial that at least one has covert bits encoded based on the angle from the most recent snapshot. Otherwise, substitution errors can occur. Lost user commands can never cause deletions or insertions, because Alice always knows from the snapshots whether any bits were sent.

Loss of snapshots is worse than loss of user commands. If the same snapshot is lost for Alice and Bob, there are no bit errors. However, if a snapshot is lost for either Alice or Bob, this can cause bit deletions/insertions in the lost snapshot and possible substitution errors in the following snapshot.

7.6.5 Reliable Data Transport

To cope with the aforementioned bit errors, Zander et al. [70] developed a tailored reliable transport protocol. The basic idea of the protocol is that Alice explicitly lets Bob know whether she can see him or not, and in the same way Bob informs Alice.

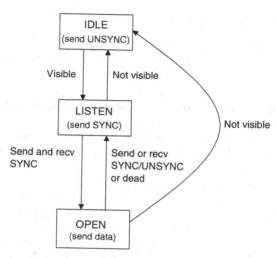

Figure 7.13. Reliable data transport state machine. (Reproduced from [70] with permission of IEEE.)

This is implemented with the help of two special channel symbols.[9] If Alice sends an UNSYNC symbol to Bob, she indicates that she cannot see Bob. If Alice sends a SYNC symbol to Bob, she indicates that she can see Bob.[10]

Alice and Bob implement the state machine shown in Figure 7.13. Initially Alice and Bob are in IDLE state. A peer in IDLE state sends UNSYNC symbols. When a peer in IDLE state sees the other peer it goes into LISTEN state and starts sending SYNC symbols. The channel's state changes to OPEN only when both peers send a SYNC to each other in the same snapshot, because only then both peers can be sure that they can see each other. Covert data is only exchanged when the channel is OPEN. An OPEN channel changes to LISTEN state if an UNSYNC or SYNC is received, or to IDLE state if visibility is lost.

To avoid synchronization errors caused by lost snapshots, transmission periods also end when one or more snapshots were lost that are used for encoding covert data (as explained before, depending on the RTT only every m-th snapshot is used). Loss of user commands does not affect bit synchronization.

The protocol works even in the presence of substitution errors, because Alice always determines whether she has sent (UN)SYNC from the snapshots, rather than what she *intended* to send.

The full details of the reliable transport protocol are explained in [70].

---

[9] Communication channel characters that have a special meaning and do not encode data bits.

[10] A drawback of using two special symbols is the increased amplitude of the induced angle changes.

7.6.6 Achievable Throughput

Zander et al. [70] evaluated the achievable throughput of FPSCC with the reliable transport protocol depending on various factors, such as the number of encoded bits per angle change (bpa), network delay, and packet loss.

7.6.6.1 *Experimental Setup.* The experiments were carried out with a prototype implementation of FPSCC for Quake III Arena [72], and were a mix of tests in a controlled testbed and tests across the Internet. The test machines were two covert game clients, a normal game client and a game server. The covert data sent was uniform random, resembling encrypted data.

In the testbed Linux's netem was used to emulate RTTs of 25 ms, 75 ms, and 125 ms. A maximum of 125 ms was chosen since players typically aim for a maximum RTT of 100–150 ms and higher RTTs noticeably affect their gaming performance [73]. Loss rates emulated with netem were 0, 0.5, and 1% (in each direction) with a maximum loss burst of two snapshots for 0.5% and three snapshots for 1% loss rate. Previous studies showed that loss in the Internet is often smaller than 1% [74], and packet loss needs to be reasonably low for good game play [73].

7.6.6.2 *Bot Players.* Long measurements with human players are problematic. Exhaustion or change in playing style over time may introduce some bias in the results. Therefore, Zander et al. mainly used *client-side* bots as players that behave consistently and do not get tired. The bots were configured to play as human-like as possible.

However, a limited number of experiments with four human players were also carried out in order to compare the angle changes per second between bots and humans. The angle change rates for yaw are very similar (11.1 changes per second for bots versus 12.2 changes per second for humans). But for pitch there is a larger difference (4.0 changes per second for bots versus 9.6 changes per second for humans). On the flat map used in the experiments there is no need to change pitch much, but randomness in mouse movements results in significantly more pitch changes for human players. This means the throughput of FPSCC is likely higher with human players.

7.6.6.3 *Results.* Zander et al. [70] measured the throughput in bits per seconds (bits/s) and the bit error rate depending on the number of players, bits encoded per angle (bpa), RTT and packet loss rate. For each distinct parameter setting, the bots played for several hours on the same map.

The throughput of FPSCC in both directions (Alice to Bob and Bob to Alice) is very similar. Therefore, here we only show the mean values over both directions in the graphs. Figure 7.14 compares the average throughput over increasing RTT for 1 bpa and 2 bpa and two or three players with 0% packet loss. Figure 7.15 compares the average throughput over increasing loss rate for 1 bpa and 2 bpa and two or three players at an RTT of 75 ms (for all results see [75]). The error bars denote the standard deviation. The bit error rate was zero in all experiments.

The throughput ranges from 10–15 bits/s at low RTTs to 2–3 bits/s at high RTTs. The throughput also reduces with an increasing number of players, since the bots were

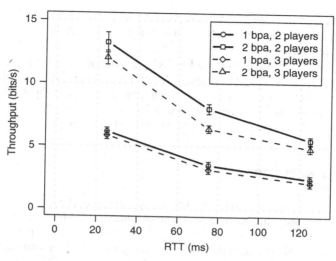

Figure 7.14. Throughput depending on Round-Trip Time (RTT), covert bits per angle change (bpa), and number of players. (Reproduced from [70] with permission of IEEE.)

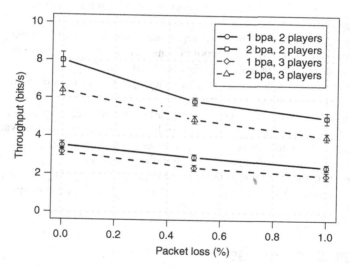

Figure 7.15. Throughput depending on packet loss rate, covert bits per angle change (bpa), and number of players. (Reproduced from [70] with permission of IEEE.)

unaware of FPSCC and did not try to optimize its throughput. However, in reality Alice and Bob could significantly improve throughput by staying in range of each other, if both are players *and* covert sender/receiver.

Zander et al. [70] also performed experiments with 1 bpa and 2 bpa encoding using two clients connected to a server via the public Internet. The average RTT between clients and server was approximately 50 ms. The throughput measured was 3.2 ± 0.3 bits/s for 1 bpa and 6.3 ± 0.2 bits/s for 2 bpa with zero bit errors, which is consistent with the testbed measurements.

The experimental results show that FPSCC with reliable transport protocol has zero bit errors. The throughput is only in the order of 10 bits/s, but this is similar to more sophisticated packet timing covert channels that provide a throughput of 6–20 bits/s [76]. Since FPS game sessions can last up to a few hours, the total amount of data that can be exchanged is substantial (in the order of tens of kilo bytes).

7.7 SOCIAL NETWORKS

Nagaraja et al. implement a covert social network botnet called *Stegobot* [77]. Stegobot consists of a botmaster and multiple bots who communicate using Facebook. The malware infects machines and then hides its data transfer within a user's uploaded images in online social networks. It intercepts Facebook image uploads to embed secret information (e.g., credit card data) into the image. The image is then transferred to other hosts (including the botmaster and the bots) via Facebook.

Stegobot is not a network steganographic approach as it applies digital image steganography to hide its messages. However, we highlight this method here because it is linked tightly to a network service.

Another approach to perform steganographic communications over Facebook was implemented by Selvi in [78]. In his proof of concept code, *FaceCat*, the author proposes to send messages encoded in ASCII via Facebook's user walls.[11] However, the approach can be considered easy to detect as FaceCat's messages appear in a completely different style as messages written by human users.

The most recent work on social network-based network steganography is *Blindspot* [79]. Blindspot's primary goal is to provide indistinguishable and unobservable anonymous communications for online protest organizers. It utilizes existing social trust relationships instead of establishing new ones and thus routes through a pre-existing social network. The tool does not change the statistic characteristics of the regular user traffic, which is achieved by piggy-backing the user's image uploads instead of uploading extra images.

7.8 INTERNET OF THINGS

The *Internet of Things* (IoT) comprises a huge variety of devices, ranging from toasters, to wearable computing equipment, to practically any Internet-connected element of a smart city (e.g., smart buildings or electric vehicles). In the remainder, we will explain information hiding in the IoT using smart buildings. The reason for this decission is

---

[11] A user wall is a part of a website where a user's posts are listed in chronological order, including uploaded images and movies.

simply that at the moment of writing there is not much research available in this emerging area and the available publications concentrate on smart buildings, which comprise a building automation network. The IoT has different areas of interest for information hiding:

1. *IoT surveillance by external source:* (Steganographic) leakage of IoT sensor data to a recipient outside of the particular IoT network. For instance, an Internet-based third party could try to perform stealthy surveillance of a smart building.
2. *Intra-organizational data leakage*: Steganographic transfer of hidden information within an IoT environment. For instance, leakage of sensitive information within a larger organizational smart building can be performed to bypass inhouse data leakage protection means between office networks.
3. *Inter-organizational data leakage*: Steganographic transfer of information from a protected LAN over an IoT environment to the Internet to bypass TCP/IP data leakage protection means of an organization. For instance, data can be leaked from an office network over a connected building automation network to the Internet.

7.8.1 From Surveillance to Steganographic Data Leakage

Being integrated into the daily life of citizens, the IoT comprises numerous sensors and actuators. Sensors measure and report events (e.g., temperature) while actuators perform actions (e.g., opening a window) and can report their state (e.g., that a window is currently opened). Many IoT environments, such as smart buildings, have their own network protocols to realize their communication. For this reason, Internet-connectivity is realized using gateways, which translate the IoT's protocol to IP and vice versa.

Given a connection to a building automation network, it is feasible to monitor events in and around buildings [80].[12] It is also feasible to record the typical behavior of inhabitants, for example, movement patterns. Surveillance should especially be considered critical when it comes to so-called *Ambient Assisted Living* (AAL). AAL technology supports handicapped persons and elders in their daily living so they can stay (longer) in their own homes before moving to a nursing home. For instance, if a person is too weak to open a window herself, AAL technology can perform the opening/closing of the window for her. In combination with bio sensors that measure sleep rhythm or blood preasure, highly sensitive information can be transferred in such building automation networks.

As shown in Figure 7.16, an attacker can obtain information at three different points of a building automation network:

---

[12] Usually a number of sensors monitor surrounding areas, for example, outdoor presence sensors are used to automatically turn on the light in a yard.

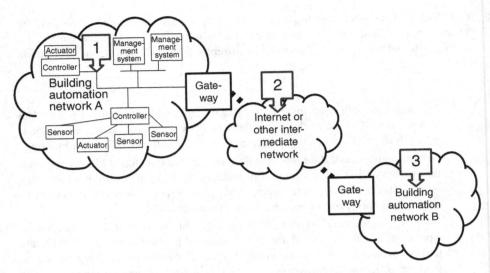

Figure 7.16. Possible locations for eavesdropper attacks and network steganographic transmissions in building automation networks.

1. If the attacker is located within the building automation network, he may have direct access to the wired or wireless transmission. Alternatively, he may have access to a controller (a so-called *direct digital control*, DDC), which is connected with the sensors and actuators, has access to the management systems (usually Windows or QNX systems), or hacked a gateway that interconnects a building with another building (probably over the Internet). In all these cases, the transfer of building automation traffic can be eavesdropped.

2. If an attacker is located on the path between two building automation networks, he can also eavesdrop the traffic exchanged. A building's intranet traffic is usually tunneled over IP when an Internet-connection is used to remotely administrate another building. In most cases, the building's network protocol is encapsulated in UDP.

3. When sensor data and control information is exchanged between different buildings, it is also feasible do directly eavesdrop information from building A inside the network of building B. Also, it is feasible to directly request sensor data of building A from the network of building B.

Network steganography can be used for bypassing traditional data leakage solutions in smart buildings. Similar to the previously introduced attack points used for eavesdropping, network steganographic traffic can be sent from a controller, a management system, a gateway, or from a remote network.

In some buildings, an inhouse TCP/IP network (e.g., an office network) is connected to (or even directly shared with) the building automation network in order to

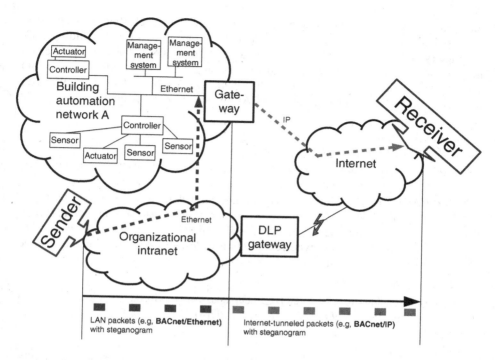

Figure 7.17. Data leakage over a building automation network to an external receiver.

use the existing network infrastructure for multiple purposes.[13] In such a case, industrial espionage can be performed using network steganography by leaking data from the office network over the building automation network's gateway (Figure 7.17). This allows bypassing DLP protection means available for TCP/IP networks [81]. In theory, all involved networks can be based on IP and a steganographic sender could directly transfer IP-based building automation packets (e.g., using the BACnet/IP protocol) to the sender. However, in most cases different network infrastructure will be used and thus, the steganogram must be carried over different low-level protocols.

7.8.2 Steganographic Communications

Whenever TCP/IP protocols are used in IoT environments—which is the case in many novel IoT environments—the creation of steganographic channels is easy as already described in Chapter 3. When it comes to legacy environments running propritary protocols or to environments running state-of-the-art non-TCP/IP protocols, the known

---

[13] Sharing is only feasible for communication protocols used in both domains, such as Ethernet. However, a gateway connection between the organizational LAN and the building automation network enables network steganography-based data leakage, too.

hiding methods must be adapted to create steganographic channels. For instance, many industrial control systems (ICS) and building automation systems (smart buildings) run proprietary and non-TCP/IP protocols, such as BACnet or KNX. The only research on network steganography published so far focuses on BACnet (a world-wide used ISO standard) but the application of known hiding methods from TCP/IP is simple. Wendzel et al. mention the following examples for BACnet-based covert channels [82]:

1. *Covert storage channel*
   - A steganogram is embedded in unused or reserved header bits (*Reserved/Unused* pattern).
   - BACnet network layer messages either embed application layer data or one of n possible *message types*. A covert storage channel can signal hidden information by selecting 1 of n possible message types for each new packet while each message type represents a different hidden information (*Value Modulation* pattern).
2. *Covert timing channel*
   - Inter-packet times between BACnet packets are modified as in case of IP-based timing channels (*Inter-packet times* pattern).

Although the majority of smart buildings are still equiped with cheap non-TCP/IP protocols (e.g., MS/TP or KNX), Ethernet and IP are on the rise. This development leads to an increasingly easy application of already known network steganographic methods to smart buildings. For instance, the previously mentioned BACnet protocol comprises an application and network layer that can be encapsulated into UDP over IP, called *BACnet/IP*. BACnet/IP is a rapidly emerging technology as an increasing number of smart buildings is being inter-connected over the Internet for two reasons. First, companies want to control all buildings, even those located on other continents, from a central location using IP as a basis, with gateway-connected subnets running older non-TCP/IP protocols such as MS/TP. Secondly, the administration of buildings can be outsourced when an IP-based remote-access is provided, redressing the need of hiring own building management operators.

Countermeasures for IoT-based steganography exist. We will discuss these countermeasures in Chapter 8.

7.9 SUMMARY

In this chapter, a wide range of diverse information hiding techniques was presented. We introduced example techniques that rely on popular Internet applications, services and communication technologies. We conclude that any area of communication networks can be subject to information hiding.

REFERENCES

1. J. Rosenberg, H. Schulzrinne, G. Camarillo, and A. Johnston. SIP: Session Initiation Protocol. IETF, RFC 3261, 2002.

2. H. Schulzrinne, S. Casner, R. Frederick, and V. Jacobson. RTP: a transport protocol for real-time applications. IETF, RFC 3550, 2003.

3. J. Lubacz, W. Mazurczyk, and K. Szczypiorski. Vice over IP. *IEEE Spectrum*, pp. 40–45, 2010.

4. W. Mazurczyk. VoIP steganography and its detection: a survey. *ACM Computing Surveys*, 46(2):20:1–20:21, December 2013.

5. W. Bender, D. Gruhl, N. Morimoto, and A. Lu. Techniques for data hiding. *IBM Systems Journal*, 35(3/4):313–336, 1996.

6. S. Zander, G. Armitage, and P. Branch. A survey of covert channels and countermeasures in computer network protocols. *Communications Surveys Tutorials, IEEE*, 9(3):44–57, Third 2007.

7. X. Wang, S. Chen, and S. Jajodia. Tracking anonymous peer-to-peer VoIP calls on the internet. In *Proceedings of the 12th ACM Conference on Computer and Communications Security (CCS '05)*, pp. 81–91. ACM, New York, NY, 2005. ACM.

8. G. Shah, A. Molina, and M. Blaze. Keyboards and covert channels. In *Proceedings of the 15th Conference on USENIX Security Symposium–Volume 15*, USENIX-SS'06, Berkeley, CA, 2006. USENIX Association.

9. G. Shah and M. Blaze. Covert channels through external interference. In *Proceedings of the 3rd USENIX Conference on Offensive Technologies*, WOOT'09, pp. 3–3, Berkeley, CA, 2009. USENIX Association.

10. W. Mazurczyk and Z. Kotulski. New VoIP traffic security scheme with digital watermarking. In J. Górski, editor, *Computer Safety, Reliability, and Security*, Vol. 4166 of *Lecture Notes in Computer Science*, pp. 170–181. Springer, Berlin, 2006.

11. W. Mazurczyk and Z. Kotulski. New security and control protocol for VoIP based on steganography and digital watermarking. In *Annales UMCS Informatica AI*, Vol. 5, pp. 417–426. Annales UMCS, 2006.

12. W. Mazurczyk and K. Szczypiorski. Covert channels in SIP for VoIP signalling. In H. Jahankhani, K. Revett, and D. Palmer-Brown, editors, *Global E-Security*, Vol. 12 of *Communications in Computer and Information Science*, pp. 65–72. Springer, Berlin, 2008.

13. W. Mazurczyk and K. Szczypiorski. Steganography of VoIP streams. In R. Meersman and Z. Tari, editors, *Proc. of the 3rd International Symposium on Information Security (IS'08), Monterrey, Mexico, Part II–Lecture Notes in Computer Science (LNCS) 5332*, pp. 1001–1018. Springer-Verlag, Berlin, November 2008.

14. P. Lloyd. An exploration of covert channels within Voice Over IP. Master's thesis, Rochester Institute of Technology, 2010.

15. L. Bai, Y. Huang, G. Hou, and B. Xiao. Covert channels based on jitter field of the RTCP header. In *IIHMSP '08 International Conference on Intelligent Information Hiding and Multimedia Signal Processing, 2008.*, pp. 1388–1391, 2008.

16. C. R. Forbes. A new covert channel over RTP. Master's thesis, Rochester Institute of Technology, 2009.

17. J. R. C. Wieser. An evaluation of VoIP covert channels in an SBC setting. In *Security in Futures Security in Change*, pp. 54–58, 2010.

18. Y. Huang, J. Yuan, M. Chen, and B. Xiao. Key distribution over the covert communication based on VoIP. *Chinese Journal of Electronics*, 20(2):357–360, 2011.

19. W. Mazurczyk, P. Szaga, and K. Szczypiorski. Using transcoding for hidden communication in IP telephony. *Multimedia Tools and Applications*, 70(3):2139–2165, 2014.

20. A. Janicki, W. Mazurczyk, and K. Szczypiorski. Evaluation of efficiency of transcoding steganography. *Journal of Homeland Security and Emergency Management*, 2014.

21. H. Tian, R. Guo, J. Lu, and Y. Chen. Implementing covert communication over voice conversations with Windows Live Messenger. *Advances in information Sciences and Service Sciences (AISS)*, 4(4):18–26, 2012.

22. J. Fridrich, M. Goljan, and R. Du. Invertible authentication watermark for JPEG images. In *Proceedings of the International Conference on Information Technology: Coding and Computing, 2001*, pp. 223–227, April 2001.

23. A. Janicki and T. Staroszczyk. Speaker recognition from coded speech using Support Vector Machines. In I. Habernal and V. Matousek, editors, *Text, Speech and Dialogue*, Vol. 6836 of *Lecture Notes in Computer Science*, pp. 291–298. Springer, Berlin, 2011.

24. W. Mazurczyk, M. Karaś, and K. Szczypiorski. SkyDe: a Skype-based steganographic method. *International Journal of Computers, Communications & Control (IJCCC)*, 8(3):389–400, 2013.

25. P. Kopiczko, W. Mazurczyk, and K. Szczypiorski. Stegtorrent: a steganographic method for the P2P file sharing service. In *Proceedings of the 2013 IEEE Security and Privacy Workshops (SPW '13)*, pp. 151–157, Washington, DC, 2013.

26. TeleGeography. International call traffic growth slows as Skype's volumes soar. Technical report, TeleGeography Report, 2012.

27. S. Molnar and M. Perenyi. On the identification and analysis of Skype traffic. *International Journal of Communication Systems*, 24(1):94–117, 2011.

28. D. Bonfiglio, M. Mellia, M. Meo, N. Ritacca, and D. Rossi. Tracking down Skype traffic. In *INFOCOM 2008. The 27th IEEE Conference on Computer Communications*. pp. 261–265, April 2008.

29. D. Bonfiglio, M. Mellia, M. Meo, D. Rossi, and P. Tofanelli. Revealing Skype traffic: when randomness plays with you. In *Proceedings of the 2007 Conference on Applications, Technologies, Architectures, and Protocols for Computer Communications (SIGCOMM '07)*, pp. 37–48. ACM, New York, NY, 2007.

30. Y.-C. Chang, K.-T. Chen, C.-C. Wu, and C.-L. Lei. Inferring speech activity from encrypted Skype traffic. In *IEEE Global Telecommunications Conference, 2008 (IEEE GLOBECOM 2008)*, pp. 1–5, November 2008.

31. E. N. C. Timberg. Skype makes chats and user data more available to police. *The Washington Post*, 2012.

32. J. Berger, A. Hellenbart, B. Weiss, S. Moller, J. Gustafsson, and G. Heikkila. Estimation of quality per call in modelled telephone conversations. In *IEEE International Conference on Acoustics, Speech and Signal Processing, 2008 (ICASSP 2008)*, pp. 4809–4812, March 2008.

33. Y. Iwano, Y. Sugita, Y. Kasahara, S. Nakazato, and K. Shirai. Difference in visual information between face to face and telephone dialogues. In *1997 IEEE International Conference on*

Acoustics, Speech, and Signal Processing, 1997 (ICASSP-97), Vol. 2, pp. 1499–1502 April 1997.

34. M. Hamdaqa and L. Tahvildari. ReLACK: a reliable VoIP steganography approach. In *2011 Fifth International Conference on Secure Software Integration and Reliability Improvement (SSIRI)*, pp. 189–197, June 2011.

35. B. Cohen. The BitTorrent protocol specification, 2008.

36. B. Inc. BitTorrent Protocol Specification v1.0, 2006.

37. A. Loewenstern. DHT protocol, 2008.

38. A. Norberg. uTorrent Transport Protocol, 2009.

39. W. Mazurczyk and P. Kopiczko. Understanding BitTorrent through real measurements. *China Communications*, 10(11):107–118, November 2013.

40. S. Shalunov, G. Hazel, J. Iyengar, and M. Kuehlewind. Low extra delay background transport (LEDBAT), July 2011.

41. D. H. Lehmer. Teaching combinatorial tricks to a computer. In *Proc. Sympos. Appl. Math. Combinatorial Analysis*, Vol. 10, pp. 179–193. Amer. Math. Soc., 1960.

42. W. Mazurczyk and L. Caviglione. Steganography in modern smartphones and mitigation techniques. *IEEE Communications Surveys Tutorials*, PP(99):1–1, 2014.

43. R. Schlegel, K. Zhang, X. yong Zhou, M. Intwala, A. Kapadia, and X. Wang. Soundcomber: a stealthy and context-aware sound trojan for smartphones. In *NDSS*. The Internet Society, 2011.

44. C. Marforio, H. Ritzdorf, A. Francillon, and S. Capkun. Analysis of the communication between colluding applications on modern smartphones. In *Proceedings of the 28th Annual Computer Security Applications Conference*, ACSAC '12, pp. 51–60. ACM, New York, NY, USA, 2012.

45. J.-F. Lalande and S. Wendzel. Hiding privacy leaks in Android applications using low-attention raising covert channels. In *2013 Eighth International Conference on Availability, Reliability and Security (ARES)*, pp. 701–710, September 2013.

46. W. Gasior and L. Yang. Network covert channels on the Android platform. In *Proceedings of the Seventh Annual Workshop on Cyber Security and Information Intelligence Research*, CSIIRW '11, pp. 61:1–61:1. ACM, New York, NY, 2011.

47. W. Gasior and L. Yang. Exploring covert channel in Android platform. In *2012 International Conference on Cyber Security (CyberSecurity)*, pp. 173–177, December 2012.

48. L. Caviglione and W. Mazurczyk. Understanding information hiding in iOS. *IEEE Computer Magazine*, 2015.

49. L. Caviglione. A first look at traffic patterns of Siri. *Transactions on Emerging Telecommunications Technologies*, 26(4):664–669, 2015 .

50. N. Lucena, G. Lewandowski, and S. J. Chapin. Covert channels in IPv6. In G. Danezis and D. Martin, editors, *Privacy Enhancing Technologies*, Vol. 3856 of *Lecture Notes in Computer Science*, pp. 147–166. Springer, Berlin, 2006.

51. W. Mazurczyk and K. Szczypiorski. Evaluation of steganographic methods for oversized IP packets. *Telecommunication Systems*, 49(2):207–217, 2012.

52. W. Fraczek, W. Mazurczyk, and K. Szczypiorski. Hiding information in a Stream Control Transmission Protocol. *Computer Communications*, 35(2):159–169, January 2012.

53. K. Szczypiorski. Hiccups: hidden communication system for corrupted networks. In *Proceedings of 10th International Multi-Conference on Advanced Computer Systems*, pp. 31–40, 2013.

54. C. Krätzer, J. Dittmann, A. Lang, and T. Kühne. Wlan steganography: a first practical review. In *Proceedings of the 8th Workshop on Multimedia and Security*, MM&Sec '06, pp. 17–22. ACM, New York, NY, 2006.

55. K. Szczypiorski. A performance analysis of hiccups–a steganographic system for wlan. In *International Conference on Multimedia Information Networking and Security, 2009 (MINES '09)*, Vol. 1, pp. 569–572, November 2009.

56. A. Najafizadeh, R. Liscano, M. V. Martin, P. Mason, and M. Salmanian. Challenges in the implementation and simulation for wireless side-channel based on intentionally corrupted {FCS}. *Procedia Computer Science*, 5(0):165–172, 2011. The 2nd International Conference on Ambient Systems, Networks and Technologies (ANT-2011) / The 8th International Conference on Mobile Web Information Systems (MobiWIS 2011).

57. M. Odor, B. Nasri, M. Salmanian, P. Mason, M. Martin, and R. Liscano. A frame handler module for a side-channel in mobile ad hoc networks. In *IEEE 34th Conference on Local Computer Networks, 2009 (LCN 2009)*, pp. 930–936, October 2009.

58. K. Szczypiorski and W. Mazurczyk. Steganography in ieee 802.11 ofdm symbols. *Security and Communication Networks*, 2011.

59. I. Grabska and K. Szczypiorski. Steganography in long term evolution systems. In *IEEE Security and Privacy Workshops*, pp. 92–99, May 2014.

60. I. Grabska and K. Szczypiorski. Steganography in wimax networks. In *2013 5th International Congress on Ultra Modern Telecommunications and Control Systems and Workshops (ICUMT)*, pp. 20–27, September 2013.

61. S. Grabski and K. Szczypiorski. Steganography in ofdm symbols of fast ieee 802.11n networks. In *2013 IEEE Security and Privacy Workshops (SPW)*, pp. 158–164, May 2013.

62. R. Holloway. *Covert DCF: A DCF-Based Covert Timing Channel In 802.11 Networks*. PhD thesis, Georgia State University, 2010.

63. T. E. Calhoun, X. Cao, Y. Li, and R. Beyah. An 802.11 mac layer covert channel. *Wireless Communications and Mobile Computing*, 12(5):393–405, 2012.

64. K. Sawicki and Z. Piotrowski. The proposal of ieee 802.11 network access point authentication mechanism using a covert channel. In *2012 19th International Conference on Microwave Radar and Wireless Communications (MIKON)*, Vol. 2, pp. 656–659, May 2012.

65. L. Frikha, Z. Trabelsi, and W. El-Hajj. Implementation of a covert channel in the 802.11 header. In *International Wireless Communications and Mobile Computing Conference, 2008 (IWCMC '08).*, pp. 594–599, August 2008.

66. R. Goncalves, M. Tummala, and J. McEachen. Analysis of a mac layer covert channel in 802.11 networks. *International Journal On Advances in Telecommunications*, 5(3/4):131–140, 2012.

67. S. Grabski and K. Szczypiorski. Network steganalysis: detection of steganography in ieee 802.11 wireless networks. In *2013 5th International Congress on Ultra Modern Telecommunications and Control Systems and Workshops (ICUMT)*, pp. 13–19, September 2013.

68. C. Krätzer, J. Dittmann, and R. Merkel. Wlan steganography revisited. In *Proc. SPIE 6819, Security, Forensics, Steganography, and Watermarking of Multimedia Contents X*, 2008.

69. S. Zander, G. Armitage, and P. Branch. Covert channels in multiplayer first person shooter online games. In *33rd Annual IEEE Conference on Local Computer Networks (LCN)*, October 2008.

70. S. Zander, G. Armitage, and P. Branch. Reliable transmission over covert channels in first person shooter multiplayer games. In *Proceedings of 34th Annual IEEE Conference on Local Computer Networks (LCN)*, October 2009.

71. G. Armitage, M. Claypool, and P. Branch. *Networking and Online Games: Understanding and Engineering Multiplayer Internet Games.* John Wiley & Sons, 2006.

72. S. Zander. CCHEF—covert channels evaluation framework, 2007. http://caia.swin.edu.au/cv/szander/cc/cchef/.

73. S. Zander and G. Armitage. Empirically measuring the QoS sensitivity of interactive online game players. In *Australian Telecommunications and Network Applications Conference (ATNAC)*, December 2004.

74. Y. Zhang, N. Duffield, V. Paxson, and S. Shenker. On the constancy of internet path properties. In *1st ACM SIGCOMM Internet Measurement Workshop*, 2001.

75. S. Zander. *Performance of selected noisy covert channels and their countermeasures in IP networks.* PhD thesis, Swinburne University of Technology, May 2010.

76. S. Gianvecchio, H. Wang, D. Wijesekera, and S. Jajodia. Model-based covert timing channels: automated modeling and evasion. In *Recent Advances in Intrusion Detection (RAID) Symposium*, September 2008.

77. S. Nagaraja, A. Houmansadr, P. Piyawongwisal, V. Singh, P. Agarwal, and N. Borisov. Stegobot: a covert social network botnet. In *Proceedings of the 13th International Conference on Information Hiding*, IH'11, pp. 299–313. Springer-Verlag, Berlin, 2011.

78. J. Selvi. Covert channels over social networks, 2012. http://www.sans.org/reading-room/whitepapers/threats/covert-channels-social-networks-33960.

79. J. Gardiner and S. Nagaraja. Blindspot: indistinguishable anonymous communications. *CoRR*, abs/1408.0784, 2014.

80. S. Wendzel. Covert and side channels in building and the prototype of a building-aware active warden. In *Proc. of the IEEE Workshop on Security and Forensics in Communication Systems (SFCS'12)*, pp. 6753–6758. June 2012.

81. S. Wendzel. *Novel approaches for network covert storage channels.* PhD thesis, University of Hagen, 2013.

82. S. Wendzel, B. Kahler, and T. Rist. Covert channels and their prevention in building automation protocols—a prototype exemplified using BACnet. In *Proceedings of the IEEE Workshop on Security of Systems and Software Resiliency (3SL'12)*, pp. 731–736. September 2012.

8

NETWORK STEGANOGRAPHY COUNTERMEASURES

The only truly secure system is one that is powered off, cast in a block of concrete and sealed in a lead-lined room with armed guards—and even then I have my doubts.

—Eugene H. Spafford

Whether network steganography is used for ''good'' or ''bad'' purposes largely depends on one's viewpoint. We do not intend to make moral judgments on the possible application scenarios for network steganography right here. However, what is clear is that in each scenario where network steganography may be used, there can be adversaries whose aim is to prevent the use of network steganography. In general, an adversary can try to eliminate the covert channel, limit its bandwidth, or monitor the use of the channel.

Before application, countermeasures must be evaluated with regard to their impact on steganographic communication. In some cases, such as observing hidden communications of criminals, it may be too risky for an adversary to raise attention and hence only countermeasures without visible influence on the covert communication can be used (mainly passive measures). Other scenarios, such as ongoing leakage of sensitive information, may require the instant elimination of a covert communication. In such cases, countermeasures noticeable by a covert channel's users can be applied.

Information Hiding in Communication Networks: Fundamentals, Mechanisms, Applications, and Countermeasures,
First Edition. Wojciech Mazurczyk, Steffen Wendzel, Sebastian Zander, Amir Houmansadr, and Krzysztof Szczypiorski.
© 2016 by The Institute of Electrical and Electronics Engineers, Inc. Published 2016 by John Wiley & Sons, Inc.

This chapter discusses the countermeasures available against network steganography. This includes measures that target a control protocol (see Chapter 4) and it also includes measures against watermarking (see Chapter 6)—a specific application of network steganography. Please note that the countermeasures against traffic type obfuscation, the other major category of information hiding in communication networks, are discussed in Chapter 5.

8.1 OVERVIEW OF COUNTERMEASURES

The first step to counter a covert channel is usually the identification of the channel, that is, becoming aware of its existence. Without knowing that a channel exists, even a naive steganographic method will be covert. Without identification, covert channels may still be detected by more general anomaly detection approaches, but to our best knowledge there has been very limited research in the anomaly detection area with respect to network steganography.

Several formal methods were developed for identifying possible covert channels in specifications or implementations of operating systems and applications during the design phase or in an already deployed system. A few formal techniques for identifying possible covert channels in network protocols also exist. Most are adaptations of the identification methods developed for operating systems or applications on single hosts. However, the identification of possible covert channels in network protocols has been largely ad hoc in the past. We discuss the identification methods for the design phase in Section 8.2.

After a network steganography technique has been identified, there are a number of available countermeasures. It may be possible to *eliminate* the use of the resulting covert channel. Some channels cannot be eliminated completely, but if one can *limit* the bandwidth of a channel severely, this will render the channel useless in practice. If one can *detect* the use of a covert channel, one can take actions against covert senders and receivers. A known detection capability may also deter potential users. To gather information about covert senders and receivers and their communication structure and patterns, one can continuously *audit* the use of covert channels.

Finally, it is important to *document* known covert channels, except channels with capacities that are too low to be significant. This makes everybody aware of their existence and potential threat, and it also deters potential users, since many steganographic techniques only provide security by obscurity.

These countermeasures are not mutually exclusive, but they can be combined. Figure 8.1 shows the existing countermeasures, their requirements, and their likely combinations.

Countermeasures can be active or passive techniques, or a combination of both. With *active* countermeasures, the adversary actively manipulates the overt traffic, the covert traffic, or both. With *passive* countermeasures, the adversary only observes the traffic. All countermeasures that eliminate or limit covert channels are active techniques. Detection methods are usually passive techniques, but detection techniques that involve manipulating control protocols also fall into the active category.

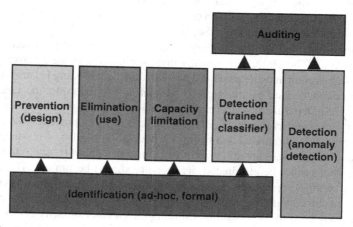

<u>Figure 8.1.</u> Countermeasures that can be used to eliminate, limit, and audit the use of network steganography.

How to deal with network steganography greatly depends on the security requirements of organizations and government policies, the applications for which the covert channels are used, and the countermeasures available in practice. While we cannot give concrete advice for specific scenarios, we will discuss a general strategy to deal with network covert channels in the following section.

8.1.1 General Strategy to Deal with Covert Channels

If a potential covert channel was not removed or cannot be removed in a protocol design or implementation, the next best option is to eliminate the use of the channel. The direct approach is to block or eliminate the use of network protocols that enable network steganography. However, blocking protocols is often impractical. Many protocols are needed to fulfill the communication needs. For particular types of network protocols, one may select protocols less prone to covert channels over more vulnerable protocols. However, often this choice is limited not only by the available protocols, but also by existing dependencies between protocols; that is, some lower layer protocols are required by higher layer protocols. Overall, there are simply too many possibilities for hiding information in network protocols.

In a completely closed network, one could of course replace or modify vulnerable protocols with secure variants, but this approach requires a lot of resources to implement and maintain. Also, it appears almost impossible to implement, except in special cases, since the operation of modern communication networks is intrinsically linked to many different protocols. Hence, currently the most common approach to prevent the use of network steganographic techniques and eliminate the resulting covert channels is *traffic normalization* (also referred to as *protocol normalization*). Traffic normalizers remove semantic ambiguities by normalizing protocol header fields or timing behaviors of network traffic passing through the normalizer. Deployed at security boundaries, they are very effective in eliminating known covert channels. We discuss the capabilities and limitations of traffic normalization in Section 8.3.

Steganographic methods that modulate message parameters are inherent in communication networks. Therefore, it is virtually impossible to eliminate all covert channels completely—an argument made previously by Moskowitz and Kang [1] and others. This fact is also acknowledged by the security standards. For example, the Orange Book released by U.S. Department of Defense treats covert channels with capacities of less than one bit per second as acceptable in many scenarios [2]. In practice, even with some vulnerable protocols eliminated and traffic normalization in place for other protocols, one must assume that there are still existing channels that could be exploited.

If a channel cannot be eliminated, its capacity should be limited. The acceptable capacity of a channel depends on the amount of information leakage that is critical. For example, if the capacity is so small that classified information cannot be leaked before it is outdated, then a channel is tolerable. Limiting the channel capacity is often difficult, because it means slowing down system or protocol mechanisms or introducing noise, which limits the performance of the system. However, some limitation techniques have been demonstrated to be very effective, especially for network steganographic timing methods that are hard to eliminate. We discuss methods to limit the capacity of covert channels in Section 8.4.

Covert channels that cannot be eliminated or limited should be audited; that is, their potential use should be monitored, which requires the capability to detect them. We discuss methods to detect network steganography in Section 8.6. Auditing also acts as deterrence to possible users and allows taking actions against actual users. If the traffic data used for auditing can be obtained before traffic normalization takes place, one may want to audit the use of all known covert channels.

8.1.2 Countermeasures Exploiting Control Protocols

As discussed in Chapter 4, often control protocols are used over covert channels to provide features, such as reliable data transport or authentication. Kaur et al. show that attacks on steganographic control protocols can be used as effective countermeasures against network steganography [3]. The attacks are based on the idea that an adversary can interact with or even manipulate the control protocol. This is often possible as control protocols do not necessarily verify the received control protocol data or only apply simple approaches to distinguish between hidden information and out-of-band data. For instance, Ping Tunnel implements a header field called ''magic number'' to verify whether a received ICMP echo request or reply packet belongs to the secret communication.

When attacking a control protocol inside a steganographic channel, Kaur et al. [3] distinguish between four different cases, which are shown in Table 8.1. If the presence of a control protocol and its header structure and functionality are known to the adversary (Case I), it is easy to detect a covert communication. However, an adversary may have no knowledge of the particular control protocol, but only knows that a control protocol-based communication is taking place, for example, due to insider information (Case II).

TABLE 8.1. Possible scenarios to attack control protocols based on [3].

Scenario	Case I	Case II	Case III	Case IV
Presence of CP	✓	✓	X	X
Understanding of CP	✓	X	✓	X

In this case, an adversary can inject random data (noise) into a covert channel, not knowing what effect the random data may cause, or he/she can try to reverse-engineer the control protocol.[1] If a control protocol and the way it is usually embedded into a carrier are known but the presence of the protocol is not detected (Case III), a blind attack can be performed in which the adversary sends disruptive commands (e.g., fake commands to terminate covert communications) to a system potentially involved in a covert communication. In the worst case (Case IV), the adversary has no information about the control protocol or its presence in the network, and no attack against the control protocol is feasible. However, it is still possible to apply known countermeasures against the covert channel itself, for example, by using traffic normalization (Section 8.3.2).

Passive countermeasures exploiting control protocols observe the communication patterns, for example, the involvement of the systems in a covert communication, the development of a covert overlay network, or the secret content itself. Active countermeasures influence the communication process, for instance,

- by redirecting messages over a separate analysis network to perform further analysis of suspicious traffic,
- by limiting the throughput of the covert channel through adversely affecting the control protocol's functionality, or
- by eliminating the covert communication (e.g., by sending spoofed termination commands using the control protocol).

In Section 8.3.3, we will highlight active attacks on control protocols and in Section 8.6.9 we will discuss passive attacks.

8.2 IDENTIFICATION AND PREVENTION DURING PROTOCOL DESIGN

Instead of being forced to eliminate, limit, or detect covert channels, at least some of them could be prevented during the design of network protocols. Eliminating semantic ambiguities, unnecessary redundancies, and unclear specifications would reduce the opportunities for network steganography. However, manual identification of possible

---

[1] No work on the reverse engineering of control protocols is available. However, botnet research produced countermeasures for *command and control* protocols and further research could show whether these countermeasures can be applied to steganographic control protocols [4].

covert channels in the protocol design is not easy and error-prone. Instead, automated formal identification methods should be used.

Several formal methods were developed for identifying covert channels in specifications or implementations of single host systems. They can identify channels during the design phase, or in an already deployed system. The existing techniques can be grouped into the following categories: information flow analysis [5,6], non-interference analysis [7], shared resource matrix (SRM) method [8,9], and covert flow tree (CFT) method [10]. Gligor [11] provides a good introduction to the different methods, except CFT.

Donaldson et al. discussed how analysis techniques, SRM in particular, could be applied to network steganography [12]. They proposed analyzing network covert channels by separately inspecting host-to-host channels on the lower network layers and intra-host channels between processes on a single host.

Hélouët et al. proposed to perform covert channel analysis for distributed systems early at the requirement level, when design decisions can still be made to eliminate or limit covert channels [13]. They argue that covert channels detected during the design phase of network protocols are not implementation-specific, and hence are likely to be present in any implementation. Their approach is based on a representation of requirements by scenarios.

Aldini and Bernardo proposed a method for combining covert channel identification and performance evaluation [14]. The advantage of this integrated approach is that it provides insights into how to trade off quality of service with covert channel capacity. They applied their methodology to the PUMP model (described in Section 8.4.4), obtaining the relation between channel capacity and the rate of served connection requests.

Overall, little work exists on formal methods for the identification of covert channels in network protocols. Furthermore, it is unclear whether the few existing methods are actually used in practice. With a lack of formal identification methods, the identification of possible covert channels remains ad hoc at best, or in the worst case is not attempted at all.

8.3 ELIMINATION OF COVERT CHANNELS

There are two complementary strategies to eliminate the use of network steganography. First, properly securing hosts and networks helps preventing the use of covert channels in the first place. Second, normalization of network traffic can eliminate a variety of steganographic techniques.

8.3.1 Securing Hosts and Networks

Securing hosts and networks can help to prevent the use of covert channels, even if the security measures are not aimed at covert channels directly.

8.3.1.1 Securing Hosts. Securing hosts connected to a network cannot remove covert channels, but it can prevent their exploitation in some application scenarios. If hosts were secured from being hacked, the installation of Trojans and modifications of

software or the network stack would be impossible. Hence, hackers or corporate spies could not exploit covert channels.

However, skilled adversaries have many options to infiltrate hosts, and detecting a hacked host is also difficult if the attacker is skilled. Furthermore, securing hosts does not address the problem in other application scenarios, where covert channels are used between hosts that are under the control of the covert sender and receiver, that is, if network steganography is used for hidden communication to plan crimes or avoid censorship.

Therefore, host security alone is insufficient and additional countermeasures must be deployed specifically against network steganography.

8.3.1.2 Securing Networks. One approach to counter tunneling channels is to block network protocols or application ports that are susceptible to covert channels. For example, ICMP is already blocked by many firewalls these days due to security concerns unrelated to covert channels, which also prevents some covert channels, such as LOKI2 [15]. Obviously, in the public Internet some protocols cannot be blocked because they are vital for the functioning of the network, such as the UDP or TCP transport protocols or the Domain Name System (DNS) protocol. Other protocols cannot be blocked, because the services they provide are too important, for example, the HTTP protocol. However, in a closed network any protocol prone to covert channels could be blocked or replaced by versions with fewer or limited covert channels.

The leakage of classified information from a high-security system to a low-security system, the classical covert channel application, can be prevented by a network design where only hosts on the same security level are allowed to communicate. Such an approach may be practical for highly secure networks, but appears impractical for diverse large open networks, such as the Internet.

Some covert channels exploit network insecurities that could be fixed. For example, bouncing covert channels [16] only work if IP source address spoofing is possible. Preventing IP address spoofing, for instance, by deploying ingress/egress filtering, would not only eliminate bouncing channels, but also solve several other network security issues. Securing computer networks against wiretapping and securing routers against compromise also prevents certain covert channels [17].

8.3.2 Traffic Normalization

Traffic normalizers (a type of active warden) can be used to eliminate steganographic methods that hide information by modifying protocols, such as altering protocol header fields, which we described in Section 3.1. Normalizers can also be used against steganographic methods that hide information in the timing of protocol messages described in Section 3.2. However, in many cases the latter is not a true normalization anymore, but it only limits the channel capacity by manipulating the timing or ordering of packets (see Section 8.4.3).[2]

---

[2] Some countermeasures that are based on exploiting control protocols or are used for removing watermarks can also be viewed as traffic normalization.

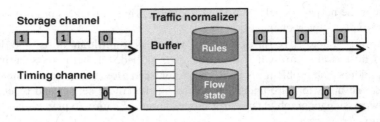

Figure 8.2. Traffic normalizers remove semantic ambiguities by modifying the content and the timing of protocol messages in order to eliminate covert channels.

Traffic normalizers remove ambiguities by converting protocol messages into their canonical forms as shown in Figure 8.2. The application of normalizers can result in side effects, since the normalization of protocol headers often includes setting header fields to default values, which means these fields are not usable anymore [18].

Several traffic normalizers exist, for example, the *network-aware active warden* [19], *Snort* [20], and *norm* [18]. Wendzel [21] compared the capabilities of several traffic normalizers. General approaches for normalizing protocols, padding, and header extensions were described by Malan et al. [22] and Fisk et al. [23]. For example, unused or reserved bits and padding can be handled easily by setting them to zero, and unknown header extensions can be removed.

Zander [24] described the normalization of covert channels, which exploit the fact that certain header fields are not always used and their use is indicated by other header fields. For example, the IP header's ID and Fragment Offset fields can be set to zero if the IP header's Don't Fragment (DF) bit is set. Lucena et al. [25] described normalization techniques specifically for the IPv6 protocol and Singh et al. [26] proposed normalization techniques to prevent ICMP tunneling. A number of other header fields can be rewritten under certain assumptions [24].

For protocol fields where all possible values are legitimate (but not necessarily used), for example, the allowed values for the IPv4 protocol field, the creation of normalization rules is more challenging and linked to constrictive side effects [23]. On the other hand, solutions for some cases exist; for example, normalizers can eliminate IP time-to-live (TTL) covert channels as described in [27] by setting the TTL value of all packets to the same value.

Traffic normalization can be performed by end hosts or by network devices, such as firewalls or proxies. The literature divides normalizers into *stateless* and *stateful* normalizers. Stateless normalizers focus on one packet at a time, and do not take previous packets into account. Stateful normalizers keep information of previously received packets to evaluate traffic based on current and historical information and can detect *more* covert channels (see [25]).

A problem of stateful normalizers is their limited flow reassembly buffer size [18]. Normalization techniques that require large buffers, for instance, to reorder network packets or to normalize the inter-arrival timings, are only useful as long as the normalizer's resources are not exhausted. Another problem in IP networks is that traffic can take different routes and thus not all packets of a flow may pass a normalizer, which results in incomplete flow information (e.g., missed packets of TCP handshakes [18]).

Traffic normalization can only be applied to all network traffic ("blind" normalization) if the normalization is transparent and does not affect the traffic significantly. Hence, blind normalization cannot eliminate all covert channels. However, if accurate detection methods exist, detected covert channels can be eliminated or limited using targeted normalization or even disruptive measures; for example, the overt traffic could simply be blocked.

8.3.2.1 Network and Transport Layers. Most normalization methods that have been proposed perform normalization on the network and transport protocol levels. They normalize certain header fields in the IP, UDP, and TCP protocols. Table 8.2 summarizes popular normalization methods that can be applied to these protocols based on [18,24,25] and [28], but it is by no means an exhaustive list of existing methods.

8.3.2.2 Application Layer. While most existing work focused on normalization techniques for network layer and transport layer protocols, the same concepts can be used for eliminating covert channels in application protocols. For example, Schear et al. proposed eliminating covert channels in HTTP responses by enforcing protocol-compliant behavior, restricting usable response headers to a fixed set in a particular order, and verifying response header fields against the corresponding object metadata and the client's request [29].

8.3.2.3 Network-Aware Traffic Normalizers. So-called *network-aware active wardens* (NAAWs) are traffic normalizers that have knowledge of the topology of a network in which they operate and implement stateful traffic inspection. NAAWs were introduced by Lewandowski et al. to normalize potential covert channels for IPv6-based traffic [28]. NAAWs are based on the concept of *active mapping* [30], which is a network protection technique that maps the network and its policies to perform improved decisions for traffic normalization. For instance, if the network topology and the routing paths are known, based on a packet's TTL a normalizer can deduce whether the packet will actually reach its destination or not, and can drop it accordingly. In [31], the idea of a NAAW is extended to a *building-aware active warden* that is capable of preventing covert channels in building automation networks (cf. Section 8.3.4).

8.3.3 Elimination Attacks Against Control Protocols

Kaur et al. [3] developed methods to counter the steganographic tools *Ping Tunnel* [32] and Smart Covert Channel Tool (SCCT) [33], which were both introduced in Section 4.1.3. Ping Tunnel is used for reliable point-to-point connections and SCCT realizes covert overlay networks with dynamic routing capability. In both cases, the key aspect of eliminating the covert communication lies in the interaction with the control protocol, for example, by performing a man-in-the-middle attack.

Fake control protocol-conforming responses can be sent to a Ping Tunnel client, which allows the observation of transferred secret information while the desired Ping

T A B L E 8.2. Well-known techniques to normalize IP, UDP, and TCP header fields and their possible side effects.

Header Field	Normalization Method	Side Effects
IP DF and More Fragments bit, Fragment Offset	Set to zero if packet is below known maximum transfer unit (MTU)	None, assuming packet is not fragmented
IPv4 ToS/Diffserv/ECN, IPv6 flow label	Set bits to zero if features unused	None, if bits really not used
IPv4 time-to-live, IPv6 hop limit	Set to a fixed value larger than longest path necessary	Higher bandwidth consumption if routing loops
IP source	Drop packet if private, localhost, broadcast address	Malformed packets are dropped
IP destination	Drop packet if destination private or nonexistent	Some packets are dropped
IP ID field	Rewrite/scramble IP ID fields	May impact diagnostics relying on increasing IDs
IPv4 options	Remove all options	May impact functionality, but IPv4 options are rarely used
IPv6 options	Many normalization techniques proposed in [28]	See [28]
Fragmented IP packets	Reassemble and refragment if necessary	None
TCP and other timestamps	Randomize low-order bits of timestamps	None, if noise introduced is low
IP, UDP, TCP packet length	If incorrect, discard or trim packets	Malformed packets are dropped
IP, UDP, TCP header length	Drop packet with header length smaller than minimum	Malformed packets are dropped
IP, UDP, TCP checksums	Drop packet if incorrect	Malformed packets are dropped
Padding in header options	Zero padding bits	None
TCP sequence and acknowledgment numbers	Rewrite initial and following sequence numbers and convert acknowledgment numbers back to original sender number space	None

Tunnel destination will not receive the hidden data. Fake replies can be combined with a passive analysis of the received covert traffic. Moreover, the magic number of traffic or the actual hidden payload can be changed by an adversary to prevent the transmission of leaked or confidential data.

Similar attacks can be performed against SCCT. It is possible to manipulate the routing table of steganographic hosts (called *agents*) in the SCCT overlay network. The adversary propagates fake updates for all routes it receives from agents, so that the adversary is represented as the shortest path to each destination.[3] Using this active countermeasure, traffic can be routed over manipulated paths in the overlay network, for example, to an analysis network. Also, SCCT's hidden payload can be modified so that the actual receiver receives only dummy data or alternatively SCCT's traffic can be blocked.

The available engineering methods for control protocols, for example, authenticating of the transferred data (cf. Chapter 4), were not considered in countermeasure research so far. Hence, the application of countermeasures is expected to become more challenging with increasingly sophisticated control protocols.

8.3.4 Covert Channel Prevention for the Internet of Things

Few specific solutions exist to counter covert communications within the Internet of Things (IoT). In these scenarios, which we describe in Section 7.8, covert channels break a security policy. The stealthiness of these channels is not of utmost importance. We discuss two countermeasures using the example of a smart building, which is an elementary component of the IoT.

In general, smart buildings are not multilevel secure; that is, their components are not assigned to separate security levels and all communication between all components is allowed. Wendzel et al. [34] proposed to design smart building networks in a way that each organizational unit within a building is handled by a separate subnet. Each subnet is assigned a security level and traffic filters are placed between the subnets. Afterwards, the Bell–LaPadula model (BLP) [35] is applied to the whole smart building network using filter rules. The BLP model mandates that a low-security entity does not read data from a high-security entity, and that a high-security entity does not write data to a low-security entity. It can be summarized as "no read up and no write down."

The traffic filters are configured so that transmissions violating the BLP model are blocked and reported. For example, the organization's management may use subnet A and interns use subnet B. Subjects with access to subnet B (e.g., direct access to a hardware device, such as a temperature sensor, or computer-based access to the network) could now try to poll sensor values of subnet A to monitor whether managers are in their office rooms. However, due to the BLP-enforcing firewalls, requests of sensor values from subnet B to A are not permitted. In [34], this BLP-based approach was realized for smart building networks running the *Building Automation and Control Networking*

---

[3] Routing tables are exchanged in form of *topology graphs* and the control protocol uses *status updates* for this purpose (cf. Section 4.3.2).

(BACnet) protocol in conjunction with the *BACnet Firewall Router* that was used as traffic filter [36].

Szlósarczyk et al. [37] and Kaur et al. [38] applied the previously introduced traffic normalization to BACnet. This allows for clearing of reserved or unused bits in the different BACnet protocol headers.

8.3.5 Removal Attacks on Network Flow Watermarking

In contrast to passive detection of watermarks (see Section 8.6.10), active attacks target both watermark invisibility and watermark robustness and mainly aim at (partially) removing watermarks.

Kiyavash et al. [39] devise active attacks on interval-based timing watermarks, known as multiflow attacks. In this approach, an attacker combines observations from multiple flows in order to detect empty timing intervals caused by watermarkers. The attacker, then, delays packets into these detected empty intervals in order to degrade watermark robustness, that is, remove the watermark. Houmansadr et al. [40] propose countermeasures to improve the robustness of interval-based watermarks against multiflow attacks. Alternatively, SWIRL, introduced in Section 6.3.4, uses flow-dependent marking to resist multiflow attacks. Lin and Hopper [41] propose another active attack in which the attackers copy marks between different flows in order to degrade watermark robustness.

8.4 LIMITING THE CHANNEL CAPACITY

As mentioned earlier, many steganographic methods that manipulate the timing of protocol messages cannot be eliminated completely. However, often their channel capacity can be drastically limited, which makes them practically ineffective. In order to determine the efficiency of capacity limitation techniques, one needs to know the capacity of an unlimited and a limited covert channel. However, capacity estimates are only known for some channels.

In the following, we first discuss briefly how to estimate the channel capacity, and mention selected work that derived capacity estimates for several channels (mainly timing channels). Then, we present the methods proposed to limit the capacity of various covert channels.

8.4.1 Channel Capacity

The channel capacity depends on the size of the object values (storage channels) or the amount of information that can be encoded in the resources (timing channels) and the rate with which the objects or resources can be modulated.

For noise-free channels, it is easy to estimate the capacity. For example, Rowland's channels [16] have a capacity of one byte per overt packet. However, the capacity in bits per second depends on the packet rate of the carrier traffic, which can vary over time.

For noisy covert channels, the capacity analysis is more difficult. Usually, the capacity is derived based on information-theoretic concepts introduced by Shannon [42]. However, it is often hard to characterize the channel noise, since it depends on the carrier traffic and other traffic in the network. .

Millen estimated the capacity of covert timing channels with noise and/or memory [43]. Moskowitz and Miller analyzed the capacity of discrete, noiseless, and memoryless timing channels [44]. Berk et al. studied the capacity of binary and multisymbol inter-packet gap timing channels [45]. Several researchers analyzed the capacity of different types of packet timing channels [46–48]. Further references for studies on the capacity of steganographic timing channels can be found in [24].

Zander estimated the capacity of covert channels embedded in the IP TTL field depending on different encoding schemes [24].

8.4.2 Limiting Address and Length Field Channels

To limit the capacity of methods that manipulate address fields, such as IP addresses (see Section 3.1), previous research suggested limiting the number of possible addresses [12,17,49]. This actually means limiting the allowed host-to-host connections, which may be possible in closed networks, but not in the open Internet. For a particular host, the sender address should always be fixed (to prevent IP source address spoofing), but the number of destination addresses or source and destination ports can hardly be limited to a small number.

Instead of limiting the interactions between hosts, sending dummy packets between random hosts inserts noise into the traffic patterns. Indirect routing achieves the same effect more efficiently [50], but still has significant overhead.

Padding all packets to a common size eliminates the packet length modulation channel [17] discussed in Section 3.1, but this adds significant overhead, especially for small packets. To increase the efficiency, Girling proposed to have a small set of available packet sizes, small enough to limit the capacity appropriately [49]. Anonymization networks use a fixed packet size to prevent traffic analysis [50], but in general the modulation of packet size cannot be effectively limited in the current IP networks.

8.4.3 Limiting Timing Channels

Multiple solutions were proposed for eliminating or at least limiting the capacity of the timing-based steganographic methods described in Section 3.2. Either random noise is introduced to mask the covert channel, or the overt channel is forced to use fixed packet or message rates, and dummy packets or messages are inserted during idle times. Wei-Ming's fuzzy-time proposal makes all clocks in a system noisy [51]. This means a covert sender cannot exactly time outgoing packets, and a covert receiver cannot accurately measure the timing of observed packets.

The so-called *link padding* method forces a packet flow to adhere to a specific traffic pattern (e.g., packet rate) by delaying packets and injecting dummy packets if necessary, which should eliminate packet timing channels [52]. However, Graham et al. showed that even if link padding is used, information about the source's traffic rate is still leaked. This is because of the inability of a padding gateway to completely isolate the processing of outgoing packets from the interrupt processing necessary to handle incoming packets [53]. These imperfections can still be used as a covert channel.

Because padding links to a single packet rate creates significant overhead, Girling proposed that senders could emit a small number of different packet rates [49]. This increases efficiency and limits covert channels to acceptable capacities. However, even this more efficient approach is impractical in public IP networks.

Message sequence timing channels can be eliminated by buffering and delaying connection attempts or service requests. Spurious data can be inserted into the network against wiretapping covert receivers, but this does not help against covert receivers on end hosts. Schear et al. proposed delaying HTTP responses to limit the capacity of HTTP-based timing channels [29]. Their proposed technique can be applied to other application protocol channels.

Wang et al. proposed to eliminate interpacket timing channels by deploying a stateful traffic controller that randomizes interpacket times on a per-flow basis [54]. They show that the controller is quite effective, since it causes very high bit error rates on covert channels. However, the controller also drastically reduces the achievable throughput for both UDP and TCP flows. Furthermore, the impact of the controller on actual application traffic remains unclear. Liu et al. also showed how the capacity of their proposed packet timing channel can be limited by introducing artificial network jitter [55]. However, they did not investigate the effects of the jitter on the performance of the overt protocols or applications.

8.4.4 Limiting Acknowledgment Message Timing Channels

A simple common security policy is the BLP model [35]. In this model, a low-security entity (Low) must not read from a high-security entity (High) and High must not write data to Low. A problem arises when Low wants to send data to High *reliably*. Reliable communication requires High to send acknowledgments (ACKs) for the data received, so that Low can resend lost data. The timing of the ACKs can be manipulated to transmit covert data. A number of methods were proposed for minimizing the capacity of this covert timing channel [56].

In the Store and Forward Protocol (SAFP), a gateway sits between Low, and High. When the gateway receives a packet from Low, it stores it in a buffer and sends an ACK to Low. Then it transmits the packet to High and waits for an ACK. When the ACK is received, the gateway removes the packet from the buffer. However, when the buffer is full, the gateway must wait for High to acknowledge a received packet until another packet from Low can be acknowledged and stored. The time it takes the gateway to send an ACK to Low is directly related to the time of receiving an ACK from High. Hence, High can ensure that the buffer is always full and continue to exploit the covert channel.

The PUMP model reduces the channel capacity much further than the SAFP [57]. The PUMP uses a historical average of High's ACK rate as the rate of sending ACKs to Low (see Figure 8.3). For every packet from High received by the trusted high process, a moving average of High's ACK rate is updated. When the trusted low process receives a message from Low, it inserts the message into a buffer, and then sends an ACK to Low

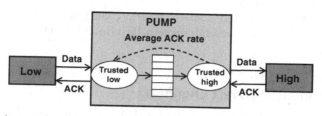

Figure 8.3. The PUMP significantly reduces covert channel capacity, because it "decouples" the high-security system's ACKs from the ACKs sent to the low-security system. (Reproduced from [58] with permission of IEEE.)

after a delay. The delay is a random variable chosen from an exponential distribution with the mean being the current average of High's ACK rate. Although the PUMP does not completely eliminate the covert channel, it significantly decreases its capacity.

While the PUMP can be and has been deployed in high-security networks, such as military networks, a widespread deployment appears unlikely. Also, the performance impact of the PUMP on high-throughput TCP flows has not been fully investigated until now.

8.4.5 Limiting Protocol Switching Channels

Wendzel and Keller presented a so-called *protocol channel-aware active warden* (PCAW) [59]. The PCAW is capable of limiting the capacity of protocol switching covert channels (cf. Section 4.2). The PCAW monitors the protocol switches occurring in each host's traffic that passes the PCAW and caches information about the last protocols used. When the PCAW sees two consecutive packets of different protocols, the second packet is slightly delayed by a time d before being forwarded. Effectively, the PCAW scrambles the protocol switches and thus destroys the hidden information as long as the covert channel's protocol switches occur fast enough. In order to be able to transfer error-free covert information, the covert channel must limit the number of protocol switches per second. In real-world network environments, a PCAW introduces slight delays in the network traffic, but the drawbacks for legitimate traffic can be minimized using a number of techniques [60].

For illustration, we assume that a protocol switching covert channel uses ICMP to signal a ''1'' bit and UDP to signal a ''0'' bit. The covert sender transfers the message ''10011'' (the sequence ICMP, UDP, UDP, ICMP, ICMP) to the receiver with minimal delay n between packets ($n \ll d$). The first ICMP packet is forwarded by the PCAW without delay. When the UDP packet reaches the PCAW, it delays the packet for a time d before forwarding it. The second UDP packet is forwarded without delay as the previous packet is also UDP. The following ICMP packet is also delayed for d and the last ICMP packet is forwarded without delay due to the previously received ICMP packet. Finally, the delayed UDP packet and the delayed ICMP packet are forwarded. Hence, the receiver will see the sequence ICMP, UDP, ICMP, UDP, ICMP (or 10**101** instead of 10**011**). If the covert sender applies a coding that reduces the number of protocol switches per bit of covert information, the effect of the PCAW can be reduced.

8.4.6 Limiting RTP Covert Channels

In Section 3.3, we described a steganographic method that encodes hidden information in late RTP packets that are discarded by the overt receiver (because they are late), but can be decoded by the covert receiver. Rezaei et al. [61] presented a method to disrupt this covert channel. In their approach, a warden inspects packets before they leave the covert sender's network or before they reach any covert receiver. For each RTP flow, the warden tracks the sequence numbers of the RTP packets, decides which packets are late, and then discards the late packets. In order to decide which packets are late, the warden must have a reasonably good estimate of the size of the overt receiver's buffer (which can be difficult to obtain in practice). If the buffer size estimate is good, Rezaei et al. show that their proposed method can detect late covert packets with a rate of more than 98%. Hence, the approach successfully disrupts the covert channel without any impact on the quality of the RTP stream.

8.5 GENERAL DETECTION TECHNIQUES AND METRICS

The detection of covert channels is commonly based on statistical approaches or machine learning (ML) techniques. In both cases, the behavior of actual observed network traffic is compared against known assumed behavior of normal traffic and (if known) the assumed behavior of carrier traffic with covert channels. The behavior is measured in the form of characteristics (also called *features* in ML terminology) computed for the actual observed traffic and traffic previously used to determine a *decision threshold* (or *decision boundary*) between the normal traffic and the traffic with steganography (training of the detection system).

With statistical approaches, the decision boundary usually must be determined manually (by a human) during the training, and for some approaches proposed in the literature it is unclear how to choose an effective boundary. In contrast, ML methods determine the decision boundary automatically during the training phase. A disadvantage with some ML techniques is that the decision boundary can be very hard to interpret, effectively turning the detection system into a black box.

With all detection methods, it is important to avoid creating a detector that performs very well for the training data, but does not perform well for the actual observed traffic (*overfitting* on the training data).

In this section, we provide an overview about existing ML methods and detection metrics that have been used to detect a number of covert channels. In Section 8.6, we discuss the actual detection approaches for different covert channels that make use of the methods discussed in this section.

8.5.1 Machine Learning

Supervised ML techniques build a classifier in the training phase based on data instances with class labels attached. In the case of network steganography, there are usually two *classes*—normal network traffic and network traffic with covert channels.

The classifier is build so that the data instances are "optimally" separated into the different classes based on the characteristics/features of the instances other than the class label. All better ML techniques build "optimal" classifiers, but avoid making the classifier too specific (avoid overfitting). The trained classifier is then used to classify data instances of an unknown class.

Figure 8.4 depicts the use of supervised ML or supervised anomaly detection approaches. First, a classifier is trained based on the characteristics (features) of examples of normal traffic and traffic with covert channels. Then this classifier can be applied to detect covert channels in the inspected traffic. There are also semi-supervised techniques that only require training on the normal traffic.

Unsupervised (clustering) techniques or unsupervised anomaly detection methods are not trained. Instead, they can automatically cluster the data or detect anomalous data instances in the observed traffic. Nevertheless, sometimes they still require some comparison of the detected clusters or the anomalous instances with the characteristics of normal traffic to identify possible covert channels.

Figure 8.5 shows the approach of using an unsupervised method. First, the network data are clustered or abnormal traffic is identified based on the characteristics (features) of traffic. Then the characteristics of the clusters or anomalies are inspected to detect covert channels.

Supervised ML techniques are often more accurate, but also less robust. Without proper retraining, they can easily fail if the characteristics of one of the classes change. Unsupervised techniques are often less accurate, but are more robust against changes in traffic characteristics.

There are many different ML algorithms. Previous research showed that for classification of network traffic the best techniques provide similar accuracy, but differ significantly in training time and classification speed [62].

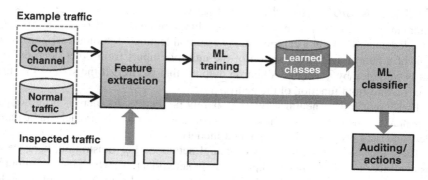

Figure 8.4. Network steganography detection with supervised ML techniques.

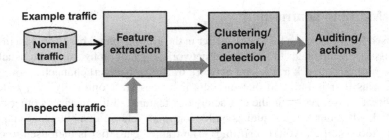

Figure 8.5. Network steganography detection with unsupervised ML techniques.

In the following sections, we briefly describe three different ML techniques that have been successfully used to detect covert channels.

8.5.1.1 Decision Trees. A decision tree model can be presented in the form of a tree with many nodes connected by branches. The top node is called the root node. A decision tree grows from the root node, splitting the data instances at each level to form new nodes. Nodes that are at the end of branches are called leaf nodes. Each node in a decision tree represents a feature test, and the branches represent possible answers. Each leaf node belongs to one class.

There is no predetermined limit to the number of levels a tree can have. The complexity of a tree, as measured by the depth and breadth, increases as the number of independent features increases.

Using a decision tree algorithm has the advantage that the model is predictive and descriptive. A human can interpret the resulting classifier (the classification tree), although with increasing size this becomes difficult.

Many different decision tree classifiers exist, but some of the basic algorithms do not provide high accuracy. The C4.5 decision tree classifier [63] performs well and has been used by a number of researchers to detect covert channels.

8.5.1.2 Neural Networks. Neural networks are based on an early model of the human brain function. The basic building block of neural networks is a processing unit called a neuron, which captures many essential features of biological neurons. The output of a neuron is a combination of the multiple inputs from other neurons. Each input is weighted by a weight. A neuron outputs, that is, fires, if the sum of the inputs exceeds a threshold function of the neuron.

The output from a neural network is purely predictive. As there is no descriptive component to a neural network model, the resulting classification model can be hard to understand (and is essentially a black box model).

There are many different types of neural networks. For illustration, here we will consider only the well-known backpropagation networks [64], which have an input and an output layer of neurons, and between these two layers have one or multiple hidden layers of neurons. Figure 8.6 shows a backpropagation network.

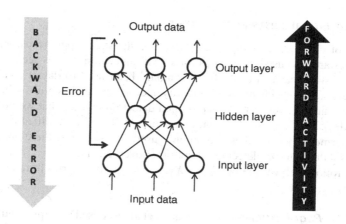

Figure 8.6. Backpropagation neural network.

A backpropagation network trains with a two-step procedure. The activity from the input pattern flows forward through the network, and the error signal flows backward to adjust the weights. The generalized delta rule allows for an adjustment of weights leading to the hidden layer neurons in addition to the usual adjustments to the weights leading to the output layer neurons. Using the generalized delta rule effectively "backpropagates" the error adjustment.

8.5.1.3 *Support Vector Machines.* The basic idea behind support vector machines (SVMs) is to map certain training data (input space) into a higher dimensional space (feature space) in order to provide a hyperplane, which can be used to separate and thus classify an input data set [65]. As evaluating the hyperplane in high-dimensional space can be computationally expensive, kernel functions are used to perform evaluations directly in the nonlinear input space.

Training data from the input space provided to the SVM must be translated into vectors of real or binary numbers. Once the data are mapped into the feature space, one of a number of potential hyperplane functions (separating the data set) could be selected. Of these, an optimal hyperplane must be chosen. The optimal hyperplane is the plane that provides the largest margin of separation between the two data sets (normal traffic and covert channels) and hence is considered most likely to perform accurate classification.

A subset of the training data that is on the edge of the margin of separation is known as support vectors. These are the values that are used to train the optimal hyperplane. To provide better generalization, values close to the hyperplane may be ignored to provide a larger margin of separation. The selection of appropriate support vectors can be done using various methods such as cross-validation.

The result of the learning process is a vector of feature weights that acts as a classifier. When an input vector is combined with the feature weights vector, the confidence of the input belonging to a trained class is returned.

8.5.2 Statistical Tests and Metrics

A number of detection techniques are based on statistical approaches, and often the test statistics are also used as features for an ML algorithm. In this section, we briefly describe statistic measures that have commonly been used to detect covert channels (mostly packet timing channels): regularity metric, the Kolmogorov–Smirnov (KS) test, entropy, and entropy rate. Other applied detection metrics that we do not discuss here include Welch's t-test [66], autocorrelation [67], and a modified chi-square test [68]. Furthermore, besides these commonly used test statistics, there are many other approaches tailored to the detection of specific covert channels.

In the following, we suppose $X = \left[X_1, \ldots, X_n\right]$ to be a series of random variables with values $x_1, \ldots, x_n$.

8.5.2.1 Regularity Measure.
Cabuk et al. proposed a simple regularity metric to detect interpacket gap timing channels [69]. Their metric determines whether the delays between packets are relatively constant or not.

Let w be a window of packets, and let σ_W be the standard deviation of the interpacket delays for window w. The regularity is defined as the standard deviation of the normalized pairwise differences between all σ:

$$R = \text{stdev}\left(\frac{\left|\sigma_i - \sigma_j\right|}{\sigma_i}, \forall i, j < i\right). \tag{8.1}$$

The regularity metric was developed to detect channels that map covert bits to different delay values. For example, for a binary channel there are two distinct delay values, one for a zero bit and another for a one bit. It does not work for more complex encoding schemes where each bit value may be mapped to several distinct delay values.

8.5.2.2 Kolmogorov–Smirnov Test.
The two-sample KS test verifies the hypothesis that two samples were drawn from the same distribution. A low KS test statistic means that the distributions are similar, whereas a high KS test statistic means the distributions are different. The KS test is distribution-free, meaning that it is applicable to a variety of types of data with different distributions.

Let $F(x)$ be the empirical cumulative distribution function of X. The KS test statistic for two empirical distribution functions F_1 and F_2 is

$$D_{KS} = \sup_x \left|F_1(x) - F_2(x)\right|, \tag{8.2}$$

where $\sup_x$ is the supremum of the set of distances.

The KS test computes the test statistic D_{KS} for two distributions. However, for covert channel detection with ML, one needs a feature that reflects how different a distribution under investigation is from the sets of distributions characterizing normal network traffic and/or covert channel traffic. Zander [24] computed the set of KS test statistics between the interpacket time distribution of one particular traffic flow (covert or normal) and the interpacket time distributions of all normal traffic flows, and then used the mean of the set of KS statistics as feature to train an ML classifier to detect interpacket time covert channels.

8.5.2.3 Kullback–Leibler Distance/Divergence.
The Kullback–Leibler (KL) distance (or divergence) is a nonsymmetric measure of the difference between two probability distributions. For two discrete probability distributions $P = \{p_1, \ldots, p_n\}$ and $Q = \{q_1, \ldots, q_n\}$, it is defined as

$$D_{KL}(P, Q) = \sum_i p_i \ln \frac{p_i}{q_i}. \tag{8.3}$$

D_{KL} is always equal to or greater than zero and it increases with increasing difference of P and Q.

8.5.2.4 Entropy.
The first-order entropy (also called Shannon entropy) is a useful metric to compare the shape of distributions of random variables. High entropy values indicate higher uncertainty, whereas low entropy values indicate lower uncertainty of the random variables. The entropy is defined as [70][4]

$$EN(X_1, \ldots, X_m) = -\sum P(x_1, \ldots, x_m) \cdot \log_2 P(x_1, \ldots, x_m). \tag{8.4}$$

Whether low or high entropy indicates a covert channel depends on the encoding method. For example, for methods that encode covert bits in unused header fields, normal traffic shows very low entropy, but the entropy for covert channels is high. In contrast, the entropy for interpacket gaps of normal traffic is usually high, whereas for simple covert channels encoded in certain fixed interpacket gaps the entropy is low.

8.5.2.5 Entropy Rate.
For comparing the complexity/regularity of a time series of random variables, we can compute an estimate of the entropy rate. The entropy rate is high if the data series has high complexity, and it is low if the series has regularities.

The exact entropy rate of a finite sequence of observations cannot be measured and must be estimated. The corrected conditional entropy (CCE) can be used to estimate the entropy rate [71]. The conditional entropy (CE) of X_m given the previously observed sequence $X_1, \ldots, X_{m-1}$ is

$$CE(X_m | X_1, \ldots, X_{m-1}) = EN(X_1, \ldots, X_m) - EN(X_1, \ldots, X_{m-1}). \tag{8.5}$$

---

[4] Usually, the detector computes an estimate of the entropy, since the detector can only estimate the probabilities from the observed distribution (histogram).

Then the CCE is defined as

$$\text{CCE}\left(X_m|X_1,\ldots,X_{m-1}\right) = \text{CE}\left(X_m|X_1,\ldots,X_{m-1}\right) + \text{perc}\left(X_m\right)\text{EN}\left(X_1\right), \quad (8.6)$$

where $\text{perc}\left(X_m\right)$ is the percentage of unique patterns of length m and $\text{EN}\left(X_1\right)$ is the first-order entropy of X_1. The estimate of the entropy rate is the minimum of CCE over different values of m:

$$\text{EER}\left(X\right) = \min\left(\text{CCE}\left(X,m\right)|m=1,\ldots,n\right). \quad (8.7)$$

The minimum exists because CE decreases while the corrective term perc($\cdot$) increases with increasing m.

To compute the entropy rate, the data need to be binned. Previous research found that equiprobable binning[5] of the data was very effective [71,72]. The number of bins Q must be chosen *a priori*. A larger Q retains a larger amount of information. However, if Q is too large, the number of possible patterns is increased exponentially (Q^m) and the ability to recognize longer patterns is reduced. The common approach is to select an "optimal" Q based on initial experiments.

Covert channels typically introduce higher complexity, so the entropy rate will be higher for covert channels than for normal traffic.

8.6 DETECTION TECHNIQUES FOR COVERT CHANNELS

Reliable detection techniques are a prerequisite for the auditing of covert channels. As explained in the previous section, all proposed detection methods for covert channels are based on the detection of nonstandard or abnormal behavior. It is assumed that the warden knows the normal behavior of network protocols and hosts and can detect deviating "abnormal" behavior caused by covert channels.

Many of the covert channels described in Chapter 3 only provide security by obscurity, and are easy to detect. However, channels that look identical to normal protocols are hard to detect. For example, the TCP ISN channel proposed in [73] has a value distribution matching the distribution of real operating systems. The only way a warden could reliably detect these covert channels is by somehow detecting the embedding process at the covert sender.

Furthermore, it is difficult to detect covert channels if there is a lot of variation in the normal behavior of a protocol or protocol mechanism. For example, Krätzer's covert channel that is based on frame duplication in WLANs [74] is potentially hard to detect, because depending on the conditions frame retransmission rates vary significantly between different WLANs.

---

[5] The size of bins is selected so that each bin holds approximately the same number of instances.

In the following, we describe the approaches proposed to detect different types of covert channels.

8.6.1 General Detection Approaches

Smith and Knight derived a general method for detecting storage or timing covert channels based on statistical inference techniques [75]. The probability of traffic being a covert channel is estimated based on deviations between the value distributions of characteristics of the traffic under investigation and regular traffic. However, it is not clear how well this approach works with different network covert channels in practice. The more specific approaches described in the following section are likely more accurate, since they are tailored to the specific covert channels.

8.6.2 Header Field Channels

Most protocol standards mandate that unused or reserved bits and padding must be filled with specific values (e.g., zero bits). Even if standards do not mandate specific values, the values used by actual implementations can be viewed as *de facto* standards [73]. All covert channels based on nonstandard use of protocols are easy to detect. Furthermore, some proposed covert channels are obsolete now because previously unused bits are now used (e.g., some bits in the IP header are now widely used for explicit congestion notification), or defined messages or extension headers are not used anymore in current networks and their presence would be suspicious (e.g., IP timestamp header extensions).

Other covert channels exploit the fact that some header fields have arbitrary values within the requirements of the standard. However, if the fields are naïvely used and the resulting value distributions are different from the typical distributions generated by operating systems, the covert channels are easy to detect [73]. Either a classifier is trained on the normal and abnormal behavior, or a classifier is trained on the normal behavior and used to detect anomalies.

Sohn et al. demonstrated that covert channels with a simple encoding hidden in the IP ID or the initial sequence number (ISN) of TCP (proposed by Rowland [16]) are discovered with high accuracy by SVMs [76]. They evaluated different feature sets and achieved classification accuracies of up to 99%.

Zhao and Shi [77] proposed a method to detect ISN channels based on phase space reconstruction. A statistical model for legitimate ISNs is constructed based on the high-order statistic analysis in the reconstructed phase space. The model can then be used to detect abnormal ISNs caused by covert channels. In contrast to the SVM-based technique, this approach only requires training data for legitimate ISNs. It also has reduced computational complexity, which makes it more suitable for online detection. Simulation-based results show that the new approach also outperforms the SVM-based method in terms of accuracy and speed.

Tumoian and Anikeev analyzed the accuracy of a neural network to detect TCP ISN covert channels [78]. First, the neural network was trained to predict successive ISNs for different operating systems. Then, the ISNs used by the hosts were monitored and compared with the prediction models. An actual ISN sequence not matching any model indicates a covert channel. Tumoian and Anikeev found that for more than 100 consecutive ISNs observed the detection accuracy reaches 99%.

Zander showed that C4.5 decision tree classifiers can be used to detect covert channels in the IP TTL field [24]. If the warden is only one hop away from the covert sender, detection is trivial as any observed TTL variation reveals the covert channel. Even if the warden is further away, the results in [24] show that the TTL covert channel is detected with over 95% accuracy, if a large fraction of overt packets is used. Even if only a small fraction is used, the accuracy is still 85–90%, as the TTL covert channel differs from normal TTL variation.

Application protocol header field covert channels can be detected in similar ways, as discussed for HTTP in [29].

Zhiyong et al. [68] showed that a modified Pearson chi-square test can be used to detect the packet size modulation method described in Section 3.1.1 with high probability, if the covert sender chooses a high covert channel capacity. However, if the covert sender deliberately reduces the channel capacity, the covert traffic cannot be distinguished from the normal traffic anymore.

Zhai et al. [79] proposed a method to detect covert channels in the TCP flags field using Markov models that model the TCP state changes from observed traffic. A baseline Markov model is fitted on normal traffic. The baseline model can then be compared with models built from normal traffic or traffic with covert channels using the KL distance (or divergence), whose value is growing with increasing difference between the probability distributions of two compared models. The results show that the KL metric can distinguish covert channels from normal traffic if the embedding ratio (number of covert to normal bits) is high.

Liu et al. [80] proposed a real-time covert channel detection system. The system runs on a secure virtual machine that mimics an existing vulnerable virtual machine. Any differences between the two virtual machines can be identified in real time. Unlike other detection systems, the warden does not require historical data to construct a model. Liu et al. demonstrated that in experiments the warden can detect covert channels with a high success rate and low latency.

8.6.3 Timestamp Channels

Hintz proposed a detection method for covert channels encoded in the TCP timestamp option [81]. In low-speed networks, a randomness test can be applied to the least significant bit (LSB) of the timestamps. Too much randomness reveals the presence of a covert channel.

In high-speed networks, the segment rate is usually larger than the TCP clock's tick rate, which is only between 1 Hz and 1 kHz. A warden can detect the channel by computing the ratio of different timestamps used and the total number of possible timestamps (depending on the duration of the connection). For a normal connection the ratio should be close to 1 (at least one segment sent every clock tick), but for the covert channel it is close to 0.5, since if a timestamp's LSB is not equal to the covert bit to be sent, one clock tick is skipped.

8.6.4 Packet Rate and Timing Channels

Venkatraman and Newman-Wolfe proposed to audit the change of traffic rates over time to detect packet rate channels [82]. If the traffic rate of one host changes by more than a certain threshold, this could indicate a covert channel. They proposed setting the threshold to the standard deviation of the regular rate change observed in the past for a large set of hosts.

Cabuk et al. proposed the regularity metric introduced in Section 8.5.2.1 to detect on/off packet rate timing channels [69]. They showed that their technique detects covert channels even if the sender changes the packet rates or there is random noise. Cabuk et al. also evaluated a technique to detect interpacket delay covert channels based on their compressibility [83]. In this approach, a series of recorded interpacket delays are converted to strings. The strings are compressed and the compressibility of a string is used to detect the presence of a covert channel. The method exploits the fact that simple interpacket gap covert channels generate traffic with a few characteristic interarrival times to signal the different covert bits and thus result in strings that can be compressed more efficiently than strings for normal random interpacket delays. Cabuk et al. also presented a detection approach based on the calculation of an ϵ-similarity metric they defined [69].[6]

Berk et al. proposed methods for detecting simple binary or multisymbol interpacket gap timing channels [45]. For binary channels, the interpacket time histogram has two distinct spikes, and the mean is between the spikes and has a very low frequency. For normal flows, the histogram has a higher frequency at the mean. This difference reveals covert channels.

For multisymbol channels, Berk et al. argued that a skilled covert sender would pick a symbol distribution that maximizes the capacity. The warden can also estimate the optimal symbol distribution, compare it with the distribution of the traffic under observation using a similarity test, and detect the presence of the covert channel if both distributions are similar. However, a covert sender would likely not choose to maximize the capacity if this compromises the channel. A practical problem is that the warden would have to build the channel matrix for each suspect traffic flow or have a very large number of pre-built channel matrices [45].

Gianvecchio and Wang proposed entropy-based metrics to identify different interpacket gap timing channels [72]. They showed that previous metrics (simple regularity test and KS test) are unable to identify all channels, but a combination of entropy and estimated entropy rate detects all channels.

Stillman proposed to detect timing channels by computing plausible covert bit strings from the interpacket gaps and scanning for these bit strings in the sender's random access memory [84]. The approach requires that the warden has access to the memory of the covert sender. Also, in practice it is very difficult to compute the bit strings without

---

[6] ϵ-similarity compares the shape of distributions. It is unclear whether ϵ-similarity provides an advantage over entropy.

knowledge of the secret shared by covert sender and receiver. Furthermore, identifying the bit strings in memory is hard if the covert sender uses memory obfuscation techniques.

Archibald and Ghosal [66] compared a number of metrics to detect interpacket gap timing channels and also investigated the detection accuracy of SVMs trained with multiple metrics. Their results show that detection accuracy of the SVM classifier is over 90% for long windows of training data (1000 interpacket gap values) when the classification model is periodically reparametrized.

Shrestha et al. [67] showed that CCE is not a reliable metric to detect interpacket gap timing channels when the size of the covert message is short. They propose to use autocorrelation as additional detection metric, since even with small message sizes the covert channel reduces the autocorrelation between interpacket times. This is consistent with the earlier results of Zander et al. [85] showing that if an application normally produces traffic with correlated interpacket times, the covert channel can be detected easily as it produces uncorrelated interpacket times.

8.6.5 Payload Tunneling

Sohn et al. trained SVM classifiers for detecting covert channels embedded in ICMP echo packets. They achieved classification accuracies of up to 99% when training a classifier on normal packets and abnormal packets generated by the payload tunneling tool Loki [86].

Pack et al. proposed detecting HTTP tunnels by using behavior profiles of traffic flows [87]. The profiles are based on a number of metrics, such as the average packet size, number of packets, ratio of small and large packets, change of packet size patterns, total number of packets sent/received, or connection duration. If the behavior profile of a flow deviates from normal behavior, it is likely to be an HTTP tunnel.

Borders and Prakash developed a tool for detecting covert channels over outbound HTTP tunnels based on a similar approach [88]. Their tool analyzes HTTP traffic over a training period, and then it is able to detect abnormal HTTP flows using metrics such as request size, request regularity, time between requests, time of the day, or bandwidth.

8.6.6 Channels in Multiplayer Games

In Section 7.6, we described a steganographic technique that allows "players" (in a non-MITM scenario these are actually covert senders and receivers) to covertly exchange information in multiplayer games or virtual worlds.

This channel cannot be eliminated because player movement is intrinsic to FPS games (or other games or virtual worlds). A warden could introduce noise to limit the channel's capacity, but the covert sender could always counter this with an increased amplitude of the covert signal, so long as the view angle fluctuations do not become visible. However, the channel may be detected in a number of ways by a warden that can analyze the players' movements in a live or recorded game.

Slight angle movements of another player character are basically invisible to human players. As shown in [24], players did not notice anything strange in the movements of the avatar being manipulated by the covert sender. Even the unaware player whose avatar was used by the covert sender did not notice any visual abnormalities.

The warden could instead examine the distribution of view angle changes. If the warden controls the server, a modified server could log all player movements for later analysis. Alternatively, the warden could join as a regular player and log all received player movement updates, although covert data would only be received when the covert sender's avatar is visible to the warden's avatar. While there is little difference between angle values of normal players and players acting as covert sender [24], if the warden is aware of the covert channel's encoding and compares modulo angle changes, the covert channel can be detected with 95% accuracy [24].

8.6.7 Protocol Switching Channels

In Section 4.2, we introduced protocol switching covert channels. Wendzel and Zander showed that these channels can be detected using C4.5 decision tree classifiers [89]. They build a classifier based on two metrics, the average number of protocol switches per packet and the average time between protocol switches, and trained it on traffic with covert channels and regular traffic. The results show that protocol switching channels can be detected with 98–99% accuracy for high-rate covert channels; however, the accuracy reduces to 96% if the throughput of the covert channel, that is, the number of protocol switches, is reduced.

8.6.8 VoIP-Based Channels

There is still no universal "one size fits all" countermeasure solution for VoIP steganography (for details on VoIP steganography methods, please refer to Sections 7.1 and 3.3.1). Thus, detection methods must be adjusted precisely to the specific information hiding technique.

Let us consider again the hidden communication scenarios presented in Section 2.12, where there are three possible locations for a warden (W1–W3). A node that performs steganalysis can be placed near the sender or receiver of the overt communication, or at some intermediate node. Moreover, the warden can monitor network traffic in single (centralized warden) or multiple locations (distributed warden). In general, the localization and number of locations in which the warden is able to inspect traffic greatly influence the effectiveness of the detection method.

For example, in scenario (1) from Figure 2.12, if a warden operates in an overt/covert sender's local area network (LAN), then some steganographic methods (e.g., those that utilize packet delay) are trivial to detect. This is because the anomaly introduced into the packet stream will be easily spotted near the transmitter. However, if a warden is present only in the sender's or receiver's LAN, then for scenario (4) the hidden communication will remain undiscovered.

Moreover, if a distributed warden has access to the same traffic flow in several network localizations, then the warden's effectiveness is likely to increase. Depending on the communication scenario, steganographic modification to the network traffic can be spotted by simple comparison in two distinct locations. For example, if scenario

(4) is utilized and a steganographic method that modifies the packets' payload is used, then by comparing the packet's payload in two different locations (e.g., in the sender's LAN and on some intermediate node in the external network) it is possible to uncover steganographic traffic modification. However, it must be emphasized that, in practice, a distributed warden is hard to realize, as VoIP calls often span across different administration responsibilities or even multiple international borders and are subject to different jurisdictions and legal systems.

Experiments with real-life VoIP connections were conducted by Mazurczyk et al. [90]. The results show that from VoIP timing methods only the one that introduces intentional losses is practically applicable, but it offers low steganographic bandwidth (approximately 1 bit/s). The other methods that employ reordering or modification of interpacket delays are impractical and easy to detect. Reordering of the RTP packets was never witnessed during the experiments, and interpacket delays varied so much because of the network conditions that applying successful steganographic techniques would be a difficult task and would result in a very low steganographic bandwidth (if successful at all).

With regards to VoIP hybrid methods, LACK can offer potentially high steganographic bandwidth by mimicking delay spikes, which can lead to packet drops at the receiving end. LACK can achieve this by intentionally causing RTP packet sequences that will surely lead to losses at the jitter buffer due to late packet drops or jitter buffer overflows. In this variant, LACK is hard to detect, and developing an effective steganalysis method remains future work.

Garateguy et al. [91] proposed a steganalysis method that relies on the classification of RTP packets as steganographic or non-steganographic. Their approach utilizes specialized random projection matrices that take advantage of prior knowledge about the normal traffic structure. Their approach is based on the assumption that normal traffic packets belong to a subspace of a smaller dimension (first method), or that they can be included in a convex set (second method). Experimental results showed that the subspace-based model is very simple and yields good performance (on average more than 90% correct classification), while the convex set-based method is more accurate, but more time consuming.

Arackaparambil et al. [92] analyzed how in distribution-based steganalysis[7] the length of the window in which the distribution is measured should be chosen to provide the greatest chance for success, that is, to maintain a low rate of false positives. The chosen traffic features were particular voice payload bytes and values of the RTP sequence number field. This approach was shown to be efficient in an evaluation with real-life VoIP traces and a prototype implementation of a simple steganographic method (a simplified variant of LACK). However, in order to estimate this detection method's real performance more complex VoIP-based steganographic algorithms should be chosen for evaluation.

---

[7] A distribution-based steganalysis system analyzes the distribution of selected network traffic features for steganographic clues.

A steganalysis method for TranSteg was developed based on MFCC (Mel-frequency cepstral coefficients) parameters and GMMs (Gaussian mixture models), and was tested for various overt/covert codec pairs in the worst-case scenario, that is, in a single-warden scenario with triple transcoding [93] (for details on TranSteg see Section 7.1.2). The proposed method efficiently detected some codec pairs, for example, G.711/G.726 with an average detection probability of only 94.6%, Speex7/G.729 with 89.6%, and Speex7/iLBC with 86.3%. Other codec pairs remained more resistant to detection; for example, for the pair iLBC/AMR, an average detection probability of 67% was achieved. Successful detection of TranSteg using the proposed steganalysis method requires at least two seconds of speech data to analyze.

8.6.9 Passive Attacks Against Control Protocols

Section 8.3.3 discussed active attacks against the control protocols of the tools Ping Tunnel and SCCT. Kaur et al. [3] also identified a number of passive attacks against control protocols.

Control protocol headers contain important data, such as source and destination addresses of covert senders and receivers. Such information allows obtaining information about the involvement of parties into a covert communication. In the case of Ping Tunnel, this information can even be obtained if a Ping Tunnel proxy is used. Control protocols can obfuscate the behavior of the steganographic traffic; however, neither Ping Tunnel nor SCCT implement such a feature. For example, Ping Tunnel does not normalize the timings of its packets and since different services (e.g., SSH through Ping Tunnel) generate different interarrival times than normal ping traffic, Ping Tunnel can be detected.

Similar attacks are possible against SCCT. By sniffing the traffic generated by SCCT agents, an observer can generate topology graphs of the covert overlay network using the information of the control protocol. This information can be used to monitor the growth and structure of a hidden network and the importance of particular routing nodes. The analysis of such aspects can help to identify the most important targets, for example when a law enforcement agency wants to take down the overlay network.

8.6.10 Passive Detection of Watermarks

Various types of techniques can be used by attackers whose goal is to detect watermark presence on network flows (thus compromising watermark invisibility).

Different kinds of passive attacks have been designed and evaluated in the literature. Various *statistical* tests have been proposed as passive attacks to watermarks, including key entropy [94], Kolmogorov–Smirnov test [95], mean-square autocorrelation [96], hypothesis testing [97], and different probability tests [39,41,98,99]. Others have used *information-theoretic* tests [94,95] to passively attack watermark invisibility.

8.7 FUTURE WORK

There are several covert channels that could possibly be detected, but to the best of our knowledge there is no published work that describes detection approaches for these. For example, unusually high rates of packet loss or packet reordering (see Sections 3.2.4 and 3.2.6) or frame collisions (see Section 3.2.7) could indicate potential covert channels. Similarly, covert channels that change packet length distributions can possibly be detected, since the packet length distribution of a flow with covert channel differs from that of regular traffic flows.

Current techniques to counter network covert channels usually focus on specific channels instead of more general characteristics common to multiple channels. A combination of many countermeasures is required to achieve a comprehensive protection, which is problematic in practice.

With a classification of channels into patterns as proposed in Wendzel et al. [100], countermeasures could target generic patterns instead of specific hiding techniques. Then the number of necessary countermeasures could be greatly reduced. Furthermore, if future steganographic techniques fall into one of the existing pattern categories, they could be handled with one of the existing countermeasures. However, the implementation and evaluation of pattern-based countermeasures in practice still remain future work.

8.8 SUMMARY

This chapter discussed the general techniques to counter network steganography and also described various proposals to eliminate, limit, or detect many of the covert channels described in Chapters 3, 6 and 7.

A number of channels created using storage methods can be eliminated by traffic normalization. Storage methods that cannot be eliminated and most timing methods can be countered by limiting the capacity of the covert channel. However, several of the limitation approaches have severe side effects—they significantly reduce the performance of all network traffic.

Machine learning methods have been shown to be very effective to detect covert channels—both storage and timing methods. However, many of these techniques are still tailored toward specific covert channels; that is, they only work for specific steganographic techniques, and often they also require training examples for normal traffic and covert channels. Only recently, research into more generic detection approaches that can detect patterns of covert channels has started.

Many covert channels use control protocols, which can also be exploited to eliminate, limit, or detect the covert channel.

Currently, there is a lack of techniques for the identification of network covert channels during the protocol design phase. Furthermore, to the best of our knowledge the few existing methods are rarely applied in practice.

REFERENCES

1. I. S. Moskowitz and M. H. Kang. Covert channels—here to stay? In *9th Annual Conference on Computer Assurance*, pp. 235–244, 1994.

2. U.S. Department of Defense Standard. Trusted computer system evaluation criteria. Technical Report DOD 5200.28-STD, U.S. Department of Defense, December 1985. http://csrc.nist.gov/publications/history/dod85.pdf.

3. J. Kaur, S. Wendzel, and M. Meier. Countermeasures for covert channel-internal control protocols. In Proceedings of the 10th International Conference on Availability, Reliability and Security (ARES), pp. 422–428, IEEE, 2015.

4. S. Wendzel and J. Keller. Hidden and under control–a survey and outlook on covert channel-internal control protocols. *Annals of Telecommunications (ANTE)*, 69(7):417–430, 2014.

5. D. Denning. A lattice model of secure information flow. *Communications of the ACM*, 19(5):236–243, 1976.

6. J. K. Millen. Information flow analysis of formal specifications. In *Proceedings of the IEEE Symposium on Security and Privacy*, pp. 3–8, April 1981.

7. J. A. Goguen and J. Meseguer. Security policies and security models. In *Proceedings of the IEEE Symposium on Security and Privacy*, pp. 11–20, April 1982.

8. R. A. Kemmerer. Shared resource matrix methodology: an approach to identifying storage and timing channels. *ACM Transactions on Computer Systems (TOCS)*, 1(3):256–277, 1983.

9. R. A. Kemmerer. A practical approach to identifying storage and timing channels: twenty years later. In *Proceedings of Annual Computer Security Applications Conference (ACSAC)*, pp. 109–118, December 2002.

10. R. Kemmerer and P. Porras. Covert flow trees: a visual approach to analyzing covert storage channels. *IEEE Transactions on Software Engineering*, SE-17(11):1166–1185, 1991.

11. V. Gligor. A guide to understanding covert channel analysis of trusted systems. Technical Report NCSC-TG-030, National Computer Security Center, November 1993. http://www.radium.ncsc.mil/tpep/library/rainbow/NCSC-TG-030.html.

12. A. L. Donaldson, J. McHugh, and K. A. Nyberg. Covert channels in trusted LANs. In *Proceedings of the 11th NBS/NCSC National Computer Security Conference*, pp. 226–232, October 1988.

13. L. Hélouët, C. Jard, and M. Zeitoun. Covert channels detection in protocols using scenarios. In *Proceedings of Workshop on Security Protocols Verification (SPV)*, April 2003.

14. A. Aldini and M. Bernardo. An integrated view of security analysis and performance evaluation: trading QoS with covert channel bandwidth. In *Proceedings of 23rd International Conference on Computer Safety, Reliability and Security (SAFECOMP)*, pp. 283–296, September 2004.

15. Daemon9. LOKI2: the implementation. *Phrack Magazine*, 7(51), September 1997.

16. C. H. Rowland. Covert channels in the TCP/IP protocol suite. *First Monday*, 2(5), May 1997. http://firstmonday.org/htbin/cgiwrap/bin/ojs/index.php/fm/article/-view/528/449, retrieved: February 2015.

17. M. A. Padlipsky, D. W. Snow, and P. A. Karger. Limitations of end-to-end encryption in secure computer networks. Technical Report ESD-TR-78-158, Mitre Cor-

poration, August 1978. http://stinet.dtic.mil/cgi-bin/GetTRDoc?AD=A059221&Location=U2&doc=GetTRDoc.pdf.

18. M. Handley, C. Kreibich, and V. Paxson. Network intrusion detection: evasion, traffic normalization, and end-to-end protocol semantics. In *10th USENIX Security Symposium*, August 2001.

19. G. Lewandowski, N. Lucena, and S. Chapin. Analyzing network-aware active wardens in IPv6. In *Information Hiding*, Vol. 4437 of *Lecture Notes in Computer Science*, pp. 58–77. Springer, Berlin, 2007.

20. Snort Project. Snort Users Manual 2.9.3, May 2012.

21. S. Wendzel. The problem of traffic normalization within a covert channels network environment learning phase. In *Proceedings of GI Sicherheit 2012*, pp. 149–161. GI, 2012.

22. G. R. Malan, D. Watson, F. Jahanian, and P. Howell. Transport and application protocol scrubbing. In *Proceedings of IEEE Conference on Computer Communications (INFOCOM)*, pp. 1381–1390, March 2000.

23. G. Fisk, M. Fisk, C. Papadopoulos, and J. Neil. Eliminating steganography in internet traffic with active wardens. In *Proceedings of 5th International Workshop on Information Hiding*, October 2002.

24. S. Zander. *Performance of selected noisy covert channels and their countermeasures in IP networks*. Ph.D. thesis, Swinburne University of Technology, May 2010.

25. N. B. Lucena, G. Lewandowski, and S. J. Chapin. Covert channels in IPv6. In *Proceedings of Privacy Enhancing Technologies (PET)*, pp. 147–166, May 2005.

26. A. Singh, O. Nordström, C. Lu, and A. L. M. dos Santos. Malicious ICMP tunneling: defense against the vulnerability. In *Proceedings of 8th Australasian Conference on Information Security and Privacy (ACISP)*, pp. 226–235, July 2003.

27. S. Zander, G. Armitage, and P. Branch. Covert channels in the IP time to live field. In *Proceedings of Australian Telecommunication Networks and Applications Conference (ATNAC)*, December 2006.

28. G. Lewandowski, N. B. Lucena, and S. J. Chapin. Analyzing network-aware active wardens in IPv6. In *Information Hiding*, Vol. 4437 *of Lecture Notes in Computer Science*, pp. 58–77. Springer, Berlin, 2007.

29. N. Schear, C. Kintana, Q. Zhang, and A. Vahdat. Glavlit: preventing exfiltration at wire speed. In *Proceedings of 5th Workshop on Hot Topics in Networks (HotNets)*, November 2006.

30. U. Shankar. Active mapping: resisting NIDS evasion without altering traffic. Technical Report UCB//CSD-2-03-1246, Computer Science Division, University of California, Berkeley, December 2002.

31. S. Wendzel. Covert and side channels in building and the prototype of a building-aware active warden. In *Proceedings of the Workshop on Security and Forensics in Communication Systems (SFCS'12)*, pp. 6753–6758. IEEE, June 2012.

32. D. Stødle. Ping Tunnel—for those times when everything else is blocked, September 2011. http://www.cs.uit.no/ daniels/PingTunnel/.

33. K. Szczypiorski, I. Margasinski, et al. TrustMAS: trusted communication platform for multi-agent systems. In *Proceedings of OTM 2008*, Vol. 5332 of *Lecture Notes in Computer Science*, pp. 1019–1035. Springer, 2008.

34. S. Wendzel, B. Kahler, and T. Rist. Covert channels and their prevention in building automation protocols—a prototype exemplified using BACnet. In *Proceedings of the Workshop on Security of Systems and Software Resiliency (3SL'12)*, pp. 731–736. IEEE, September 2012.

35. D. Bell and L. LaPadula. Secure computer systems: mathematical foundation. Technical Report ESD-TR-73-278, Mitre Corp, 1973.

36. D. Holmberg, J. Bender, and M. Galler. Using the BACnet firewall router. *BACnet Today (A Supplement to ASHRAE Journal)*, 48:B10–B14, 2006.

37. S. Szlósarczyk, S. Wendzel, J. Kaur, M. Meier, and F. Schubert. Towards suppressing attacks on and improving resilience of building automation systems—an approach exemplified using BACnet. In *Proceedings of Sicherheit 2014*, Vol. 228 of *Lecture Notes in Informatics*, p. 407–418. GI, March 2012.

38. J. Kaur, J. Tonejc, S. Wendzel, and M. Meier. Securing BACnet's pitfalls. In *Proceedings of ICT Systems Security and Privacy Protection*. IFIP Advances in Information and Communication Technology, Volume 455, pp 616–629, Springer, 2015.

39. N. Kiyavash, A. Houmansadr, and N. Borisov. Multi-flow attacks against network flow watermarking schemes. In P. van Oorschot, editor, *USENIX Security Symposium*, Berkeley, CA. USENIX Association, 2008.

40. A. Houmansadr, N. Kiyavash, and N. Borisov. Multi-flow attack resistant watermarks for network flows. In *34th IEEE International Conference on Acoustics, Speech and Signal Processing (ICASSP)*, 2009.

41. Z. Lin and N. Hopper. New attacks on timing-based network flow watermarks. In *USENIX Security Symposium*, pp. 381–396, 2012.

42. C. E. Shannon. A mathematical theory of communications. *Bell Systems Technical Journal*, 27(3):379–423, 1948.

43. J. K. Millen. Covert channel capacity. In *Proceedings of the IEEE Symposium on Research in Security and Privacy*, pp. 60–66, May 1987.

44. I. S. Moskowitz and A. R. Miller. Simple timing channels. In *Proceedings of IEEE Symposium on Research in Security and Privacy*, pp. 56–64, 1994.

45. V. Berk, A. Giani, and G. Cybenko. Detection of covert channel encoding in network packet delays. Technical Report TR2005-536, Department of Computer Science, Dartmouth College, November 2005. http://www.ists.dartmouth.edu/library/149.pdf.

46. S. Gianvecchio, H. Wang, D. Wijesekera, and S. Jajodia. Model-based covert timing channels: automated modeling and evasion. In *Proceedings of Recent Advances in Intrusion Detection (RAID) Symposium*, September 2008.

47. X. Luo, E. W. W. Chan, and R. K. C. Chang. TCP covert timing channels: design and detection. In *Proceedings of IEEE/IFIP International Conference on Dependable Systems and Networks (DSN)*, June 2008.

48. L. Yao, X. Zi, L. Pan, and J. Li. A study of on/off timing channel based on packet delay distribution. *Computers & Security*, 28(8):785–794.

49. C. G. Girling. Covert channels in LAN's. *IEEE Transactions on Software Engineering*, SE-13(2):292–296, 1987.

50. R. E. Newman-Wolfe and B. R. Venkatraman. High level prevention of traffic analysis. In *Proceedings of 7th Annual Computer Security Applications Conference*, pp. 102–109, December 1991.

51. H. Wei-Ming. Reducing timing channels with fuzzy time. In *Proceedings of IEEE Computer Society Symposium on Research in Security and Privacy*, pp. 8–20, May 1991.

52. B. R. Venkatraman and R. E. Newman-Wolfe. Transmission schedules to prevent traffic analysis. In *Proceedings of 9th Annual Computer Security and Applications Conference*, pp. 108–115, December 1993.

53. B. Graham, Y. Zhu, X. Fu, and R. Bettati. Using covert channels to evaluate the effectiveness of flow confidentiality measures. In *Proceedings of 11th International Conference on Parallel and Distributed Systems*, pp. 57–63, July 2005.

54. Y. Wang, P. Chen, Y. Ge, B. Mao, and L. Xie. Traffic controller: a practical approach to block network covert timing channel. In *Proceedings of International Conference on Availability, Reliability and Security*, pp. 349–354, 2009.

55. Y. Liu, D. Ghosal, F. Armknecht, A.-R. Sadeghi, S. Schulz, and S. Katzenbeisser. Hide and seek in time–robust covert timing channels. In *Proceedings of 14th European Symposium on Research in Computer Security*, September 2009.

56. N. Ogurtsov, H. Orman, R. Schroeppel, S. O'Malley, and O. Spatscheck. Experimental results of covert channel limitation in one-way communication systems. In *Proceedings of Symposium on Network and Distributed System Security (SNDSS)*, February 1997.

57. M. H. Kang and I. S. Moskowitz. A pump for rapid, reliable, secure communication. In *Proceedings of ACM Conference on Computer and Communications Security (CCS)*, pp. 119–129, 1993.

58. S. Zander, G. Armitage, and P. Branch. A survey of covert channels and countermeasures in computer network protocols. *IEEE Communications Surveys & Tutorials*, 9(3):44–57, 2007.

59. S. Wendzel and J. Keller. Design and implementation of an active warden addressing protocol switching covert channels. In *Proceedings of the 7th International Conference on Internet Monitoring and Protection (ICIMP 2012)*, pp. 1–6. IARIA, 2012.

60. S. Wendzel and J. Keller. Preventing protocol switching covert channels. *International Journal on Advances in Security*, 5(3–4):81–93, 2012.

61. F. Rezaei, M. Hempel, P. Dongming, and H. Sharif. Disrupting and preventing late-packet covert communication using sequence number tracking. In *IEEE Military Communications Conference (MILCOM)*, pp. 599–604, November 2013.

62. N. Williams, S. Zander, and G. Armitage. A preliminary performance comparison of five machine learning algorithms for practical IP traffic flow classification. *SIGCOMM Computer Communication Review*, 36(5):5–16, 2006.

63. R. Kohavi and J. R. Quinlan. *Decision-Tree Discovery*, Chapter 16.1.3, pp. 267–276. Oxford University Press, 2002.

64. R. Rojas. *Neural Networks: A Systematic Introduction*. Springer, New York, 1996.

65. N. Cristianini and J. Shawe-Taylor. *An Introduction to Support Vector Machines and Other Kernel-Based Learning Methods*. Cambridge University Press, New York, NY, 2000.

66. R. Archibald and D. Ghosal. A comparative analysis of detection metrics for covert timing channels. *Computers & Security*, 45:284–292, 2014.

67. P. L. Shrestha, M. Hempel, F. Rezaei, and H. Sharif. Leveraging statistical feature points for generalized detection of covert timing channels. In *IEEE Military Communications Conference (MILCOM)*, pp. 7–11, October 2014.

68. C. Zhiyong, S. Ying, and S. Changxiang. Detection of insertional covert channels using chi-square test. In *International Conference on Multimedia Information Networking and Security (MINES)*, pp. 432–435, November 2009.

69. S. Cabuk, C. E. Brodley, and C. Shields. IP covert timing channels: design and detection. In *Proceedings of 11th ACM Conference on Computer and Communications Security (CCS)*, pp. 178–187, October 2004.

70. T. M. Cover and J. A. Thomas. *Elements of Information Theory. Wiley Series in Telecommunications*, Wiley, 1991.

71. A. Porta, G. Baselli, D. Liberati, N. Montano, C. Cogliati, T. Gnecchi-Ruscone, A. Malliani, and S. Cerutti. Measuring regularity by means of a corrected conditional entropy in sympathetic outflow. *Biological Cybernetics*, 78(1):71–78, 1998.

72. S. Gianvecchio and H. Wang. Detecting covert timing channels: an entropy-based approach. In *Proceedings of 14th ACM Conference on Computer and Communication Security (CCS)*, November 2007.

73. S. J. Murdoch and S. Lewis. Embedding covert channels into TCP/IP. In *Proceedings of 7th Information Hiding Workshop*, June 2005.

74. C. Krätzer, J. Dittmann, A. Lang, and T. Kühne. WLAN steganography: a first practical review. In *Proceedings of 8th ACM Multimedia and Security Workshop*, September 2006.

75. R. W. Smith and G. S. Knight. Predictable design of network-based covert communication systems. In *Proceedings of the IEEE Symposium on Security and Privacy*, pp. 311–321, 2008.

76. T. Sohn, J. Seo, and J. Moon. A study on the covert channel detection of TCP/IP header using support vector machine. In *Proceedings of 5th International Conference on Information and Communications Security*, pp. 313–324, October 2003.

77. H. Zhao and Y.-Q. Shi. Detecting covert channels in computer networks based on chaos theory. *IEEE Transactions on Information Forensics and Security*, 8(2):273–282, 2013.

78. E. Tumoian and M. Anikeev. Network based detection of passive covert channels in TCP/IP. In *Proceedings of 1st IEEE LCN Workshop on Network Security*, pp. 802–809, November 2005.

79. J. Zhai, G. Liu, and Y. Dai. A covert channel detection algorithm based on TCP Markov model. In *International Conference on Multimedia Information Networking and Security (MINES)*, pp. 893–897, November 2010.

80. A. Liu, J. Chen, and L. Yang. Real-time detection of covert channels in highly virtualized environments. In *IFIP Advances in Information and Communication Technology*, pp. 151–164, 2011.

81. A. Hintz. Covert channels in TCP and IP headers, 2003. http://www.defcon.org/images/defcon-10/dc-10-presentations/dc10-hintz-covert.ppt.

82. B. R. Venkatraman and R. E. Newman-Wolfe. Capacity estimation and auditability of network covert channels. In *Proceedings of IEEE Symposium on Security and Privacy*, pp. 186–198, May 1995.

83. S. Cabuk, C. E. Brodley, and C. Shields. IP covert channel detection. *ACM Transactions on Information and System Security (TISSEC)*, 12(4):22:1–22:29, 2009.

84. R. M. Stillman. Detecting IP covert timing channels by correlating packet timing with memory content. In *Proceedings of IEEE SoutheastCon*, pp. 204–209, 2008.

85. S. Zander, G. Armitage, and P. Branch. Stealthier inter-packet timing covert channels. In *IFIP Networking*, May 9–13, 2011.

86. T. Sohn, J. Moon, S. Lee, D. H. Lee, and J. Lim. Covert channel detection in the ICMP payload using support vector machine. In *Proceedings of 18th International Symposium on Computer and Information Sciences (ISCIS)*, pp. 828–835, November 2003.

87. D. Pack, W. Streilein, S. E. Webster, and R. K. Cunningham. Detecting HTTP tunneling activities. In *Proceedings of 3rd Annual Information Assurance Workshop*, June 2002.

88. K. Borders and A. Prakash. Web tap: detecting covert web traffic. In *Proceedings of 11th ACM Conference on Computer and Communications Security (CCS)*, pp. 110–120, October 2004.

89. S. Wendzel and S. Zander. Detecting protocol switching covert channels. In *Proceedings of the 37th IEEE Conference on Local Computer Networks (LCN)*, pp. 280–283, 2012.

90. W. Mazurczyk, K. Cabaj, and K. Szczypiorski. What are suspicious VoIP delays? *Multimedia Tools and Applications*, 57(1):109–126, 2012.

91. G. Garateguy, G. Arce, and J. Pelaez. Covert channel detection in VoIP streams. In *45th Annual Conference on Information Sciences and Systems (CISS)*, pp. 1–6, March 2011.

92. C. Arackaparambil, G. Yan, S. Bratus, and A. Caglayan. On tuning the knobs of distribution-based methods for detecting VoIP covert channels. In *45th Hawaii International Conference on System Science (HICSS)*, pp. 2431–2440, January 2012.

93. A. Janicki, W. Mazurczyk, and K. Szczypiorski. Steganalysis of transcoding steganography. *Annals of Telecommunications—Annales des telecommunications*, 69(7–8):449–460, 2014.

94. A. Houmansadr and N. Borisov. SWIRL: a scalable watermark to detect correlated network flows. In *Proceedings of the Network and Distributed System Security Symposium (NDSS'11)*, 2011.

95. A. Houmansadr, N. Kiyavash, and N. Borisov. RAINBOW: a robust and invisible non-blind watermark for network flows. In *Network and Distributed System Security Symposium*, February 2009.

96. W. Jia, F. P. Tso, Z. Ling, X. Fu, D. Xuan, and W. Yu. Blind detection of spread spectrum flow watermarks. In *INFOCOM*. IEEE, 2009.

97. X. Luo, J. Zhang, R. Perdisci, and W. Lee. On the secrecy of spread-spectrum flow watermarks. In *Computer Security—ESORICS*. Springer, 2010.

98. X. Luo, P. Zhou, J. Zhang, R. Perdisci, W. Lee, and R. K. Chang. Exposing invisible timing-based traffic watermarks with BACKLIT. In *Annual Computer Security Applications Conference (ACSAC)*, pp. 197–206, 2011.

99. P. Peng, P. Ning, and D. S. Reeves. On the secrecy of timing-based active watermarking trace-back techniques. In V. Paxson and B. Pfitzmann, editors, *IEEE Symposium on Security and Privacy*, pp. 334–349. IEEE Computer Society Press, May 2006.

100. S. Wendzel, S. Zander, B. Fechner, C. Herdin: Pattern-based survey and categorization of network covert channel techniques, ACM Computing Surveys, 47(3), pp. 50:1–26, ACM, 2015.

9

CLOSING REMARKS

We hope this book underlined that network information hiding methods are a powerful toolset and a dual-use good—like any tools, information hiding methods can be used for "good" or "bad" purposes. In the closing remarks we raise some significant questions, which are related to the future relevance of network information hiding methods, how they will be potentially used by society, and how society will be able to protect itself against them.

The roots of the described methods stem from antique times. The need for sending messages, which cannot be compromised in case of interception, had motivated people to create codes or symbols that appeared innocent, but in fact had different significance than the apparent.

Modern information hiding employs various embedding techniques, but a lot of these are the result of the transfer of some previously known method into the digital domain. An interesting exception to this notion are information hiding techniques in communication networks, a family of methods that emerged with the popularization of computer network environments. This type of data hiding can be treated as evolutionary step in the development of information hiding techniques. The growing number of communication protocols, services, and computing environments offers almost unlimited opportunities for applying a whole spectrum of steganographic methods.

As we described in Section 1.6.2, hiding methods are already used to enhance the stealthiness of malware. In addition to the network-level countermeasures presented in

Information Hiding in Communication Networks: Fundamentals, Mechanisms, Applications, and Countermeasures,
First Edition. Wojciech Mazurczyk, Steffen Wendzel, Sebastian Zander, Amir Houmansadr, and Krzysztof Szczypiorski.
© 2016 by The Institute of Electrical and Electronics Engineers, Inc. Published 2016 by John Wiley & Sons, Inc.

this book, which role will host-based methods from anti-malware research play when it comes to the analysis of a malware's network hiding methods?

Illicit activities conducted in the virtual world pose a tangible threat to the society, as recent cyber warfare events show. Indisputably, information hiding has joined the arsenal of the utilized weapons, and thus it poses a large threat to information systems' security. More importantly, the matter is pressing, because steganalysis techniques are still one step behind the newest steganography methods. Currently, there is no "one size fits all" solution available to detect covert communication in our current network security defenses systems.

It should also be noted that in many current computing platforms, like smartphones, the various techniques presented in this book can be combined with each other as well as various types of sensors essentially offering nearly unlimited options for covert communication with the surrounding environment. From this perspective, the following future trends can be deduced:

- New information hiding techniques will be continually introduced, and their degree of sophistication will increase. Hence, future malware-related traffic could be harder to detect.
- Information hiding techniques can be easily incorporated into every type of malware to provide stealthy communication of control commands, the exfiltration of confidential user data, as well as communication from isolated environments or networks.
- Information hiding-capable malware can remain cloaked for a long period of time while slowly but continuously leaking sensitive user data. Thus, this type of malware must be considered as a new advanced persistent threat that must be addressed by future security systems.

A long-term solution to these trends is to consider the potential vulnerabilities enabling covert communications from the very early design phases of desktop and mobile platforms, services, and protocols. Some opportunities for covert channels could be possibly eliminated already in the design phase. Channels that cannot be eliminated in the design phase could be identified and documented, which would greatly reduce the risk of their future misuse.

For existing devices, especially smartphones, a short-term approach would require some form of ad hoc mitigation, at least for the most hazardous threats. However, this *a posteriori* approach is very difficult, because there are not yet any universal counter-measures. We hope that through this book we are raising awareness and understanding of these information hiding techniques, which will help researchers and security experts develop the necessary countermeasures. We call on the community to focus its efforts and develop steganalysis methods that could be practically deployed in a wide range of networking environments.

An important question is whether network information hiding will also gain increasing influence on or even replace current censorship circumvention systems? If yes, right now we are in the situation that the detection of many hiding methods is difficult.

However, how can the central problem that many hiding techniques do only provide security by obscurity be tackled for censorship circumvention tools, which have to be accessible for the masses? To solve this problem, improved hiding techniques that provide real security are needed. However, once such secure techniques exist, there is the risk that they will be misused for malicious purposes.

Until now, network steganography techniques mostly provide security by obscurity. Once techniques become known, countermeasures are usually developed quickly. However, new applications and protocols (especially on the application layer) and new systems still offer many new and yet unexplored opportunities for information hiding. The rise of new hiding techniques then necessitates further development of new and improved countermeasures.

How will this cyber predator-prey arms race between improving hiding methods and countermeasures continue, if government agencies put additional effort into the development of measures for detection, limitation, and prevention? To what extent will such a trend influence the surveillance of citizens? If criminals and terrorists will make increased use of hiding methods, how can their misuse be prevented or detected on a large scale, if at all? What detection accuracy will be provided by future pattern-based countermeasures?

In Chapter 1, we have already mentioned several areas to which network information hiding could contribute in the future. For instance, there are potential use cases for network flow watermarking in automated environments, such as smart buildings or in content-centric networking. However, the majority of these areas are still in their infancies today and many open questions remain. What is certain is that the cyber predator-prey arms race of developing new improved hiding methods and developing more effective countermeasures will continue in the future.

GLOSSARY

Bouncing covert channel A covert channel where the covert sender sends packets to an intermediate host with spoofed source IP address set to the covert receiver's address (or an address of a host downstream of the covert receiver). The traffic will then "bounce off" the intermediate host and eventually reach the covert receiver. This is one particular realization of an Indirect Covert Channel.

Camouflage In nature, all solutions that utilize physical shape, texture, coloration, illumination, and so on in making animals hard to spot. This causes the information about their exact location to remain ambiguous. Organisms with camouflage typically look like an element of their habitat, for example, a rock, a twig, or a leaf.

Control protocol Also *micro-protocol*. A communication protocol that is embedded into a steganographic carrier; it regulates the communication between distributed steganographic processes.

(Information hiding) Countermeasure Effort directed toward identification, prevention, disruption, discovery, or auditing of the covert communication.

Cover area Combined area of utilized subcarriers in a packet.

Covert channel A non-obvious information exchange usually achieved by hiding messages in other objects. Typically, we distinguish between a local and a network covert channel. In a networking environment, a covert channel is created using a network steganography technique.

Covert receiver (secret receiver) An entity that extracts hidden data from a hidden data carrier using some steganographic technique.

Covert sender (secret sender) An entity that embeds hidden data into a hidden data carrier using some steganographic technique.

Crypsis A term from ecology that is used to describe the abilities of organisms to effectively hide or conceal their presence in order to avoid detection/observation. The two most notable crypsis techniques are camouflage and mimicry.

Information Hiding in Communication Networks: Fundamentals, Mechanisms, Applications, and Countermeasures,
First Edition. Wojciech Mazurczyk, Steffen Wendzel, Sebastian Zander, Amir Houmansadr, and Krzysztof Szczypiorski.
© 2016 by The Institute of Electrical and Electronics Engineers, Inc. Published 2016 by John Wiley & Sons, Inc.

Cryptosystem A pair of algorithms that uses a secret key to encrypt and decrypt data. The secret key for encrypting and decrypting can be the same (symmetric cryptosystem) or different (asymmetric cryptosystem).

Cryptosystem steganography A type of steganography where features of a cryptosystem are exploited to conceal secret data. Using these techniques, a covert channel inside a cryptosystem can be created.

(Hidden) Data carrier One or more overt traffic flows that pass between a covert sender and a covert receiver (or secret sender/receiver). Typically, the carrier can be multidimensional, that is, it offers many opportunities (places) for information hiding (called subcarriers).

Digital media steganography A type of steganography where secret data are embedded into innocent-looking digital images, audio or video files, that is, digital media content is used as a hidden data carrier.

Digital steganography Different types of steganography that can be utilized in a digital environment. It incorporates steganographic techniques for digital media content, linguistic, file system, and network steganography.

Ephemeral covert channel Covert communication where the communicating parties are not exchanging any packets directly.

Filesystem steganography A type of steganography that allows to create a steganographic file system, a first solution was originally proposed by Anderson, Needham, and Shamir.

Hybrid methods A class of network steganography techniques that combine the features of timing and storage methods.

(Digital) Image steganography A type of steganography that utilizes a digital image as a hidden data carrier. Image steganography is a subtype of digital media steganography.

Indirect covert channel A covert channel where the covert sender does not send covert data directly to the covert receiver (or a destination downstream of the covert receiver). Instead, the covert sender sends the covert data to an intermediate host, which then unknowingly forward (due to the functions of the overt traffic protocol) the covert data to the covert receiver.

Information hiding in CNs Techniques that allow concealing information in communication networks and are inseparably bounded to the information transmission process.

Local covert channel A type of covert channel that allows covert communication between different processes/applications on a single machine.

LSB Least significant bit. The last/rightmost bit in a big-endian representation of an integer.

Mimicry In nature, it characterizes all the cases in which an organism's attributes are obfuscated by adopting the characteristics of another organism. It allows an animal to hide information about its own identity by impersonating something that it is not.

MLS Multilevel steganography. A construction of at least two steganographic methods that are utilized simultaneously in such a way that one method's (the upper level) network traffic serves as a carrier for the second method (the lower level).

NEL Network environment learning. Automatic discovery of techniques that can be used between two or more peers to exchange secret data. It can be incorporated into covert communication as a phase in which peers probe for available steganographic methods (from a set of known methods) and rule out methods relying on blocked and non-routed network protocols.

Network covert channel A type of covert channel that allows covert communication through a communication network.

Network flow watermarking Information hiding techniques that manipulate the traffic patterns of a network flow (e.g., the packet timings or packet sizes) in order to inject an artificial signal into that network flow—a watermark. Network flow watermarking is primarily used for linking network flows in application scenarios where packet contents are striped of all linking information. These techniques can be treated as an application of network steganography.

Network steganography Techniques that conceal information in communication networks by creating covert (steganographic) channels for hidden communication, which are inseparably bounded to the hidden data carrier and do not destroy it. Network steganography techniques can be divided into timing, storage, and hybrid methods.

Overt traffic Visible network traffic usable as hidden data carrier generated by the covert sender or another unwitting party not involved in the hidden data transfer.

(Information hiding) Pattern An abstract description of how to solve a problem in a given context. Information hiding patterns provide an abstract description of a hiding technique and are categorized in a hierarchy.

Physical steganography A type of steganography in which a physical medium is utilized as a hidden data carrier. Most notable examples of physical steganography are information hiding by tattooing human skin, using invisible ink or microdots.

Prisoner's problem A fictitious scenario involving two prisoners who want to agree on an escape plan while a warden monitors their activities. The scenario demonstrates the problem of covert communication and its detection.

Side channel A type of a covert channel where the sender unintentionally leaks information and only the receiver desires a successful communication.

Status update An approach to minimize the size of control protocol headers.

Steganalysis The counterpart of steganography. An effort directed toward discovering the presence of a secret message.

Steganographic bandwidth A feature that characterizes a network steganography method with respect to how much secret data can be send per time unit.

Steganographic cost A metric that characterizes a network steganography method with regard to the degradation of the carrier caused by the steganogram insertion. The steganographic cost depends on the type of the carrier utilized, and if it becomes excessive, it leads to easier detection of the steganographic method.

Steganographic key (Stego-key) A secret shared between the secret sender and secret receiver that drives the embedding and extraction algorithms.

Steganographic robustness A metric that characterizes a network steganography method with regard to the amount of alteration a steganogram can withstand without destroying the secret data.

Steganography The art of communicating messages in a covert manner.

Storage methods A class of network steganography methods that modify the "places" (subcarriers) in a carrier to create a storage covert channel. These techniques hide information by modifying protocol fields, such as unused bits of a header.

Subcarrier A "place" or a timing of "events" in a carrier where secret information can be hidden using a single steganographic technique. Typically, a subcarrier takes the form of a storage or a timing covert channel.

Timing methods A class of network steganography methods that modify the timing of "events" (subcarriers) in a carrier to create a timing covert channel. These techniques hide information in the timing of protocol messages or packets. Protocol-aware timing methods require the understanding of the carrier protocol as they utilize its characteristic features for hidden data exchange. Protocol-agnostic techniques can be applied blindly to the selected carrier without in-depth understanding of the utilized protocol mechanisms and their specific features.

Traffic de-identification A class of traffic type obfuscation techniques that manipulate network traffic so that the underlying network protocol is hidden.

Traffic flow A sequence of data sent from a particular source to a particular destination.

Traffic impersonation A class of traffic type obfuscation techniques that manipulate traffic so that the underlying network protocol is not only hidden but also looks like another protocol, for example, a target protocol.

Traffic type obfuscation Techniques that conceal information in communication networks by hiding the type of network traffic flows. Traffic type obfuscation methods modify the patterns and/or contents of network traffic flows so that a third party is not able to reliably identify the form of communication, that is, the network protocol.

Traffic type obfuscation does not require a hidden data carrier, as the overt transmission channel is effectively ''destroyed.''

Type of update Indicator for the type of a following header field for status update-based control protocols.

Undetectability A metric that characterizes a network steganography method with respect to the difficulty of detecting a steganogram inside a carrier. The most popular way to detect a steganogram is to analyze statistical properties of the captured data and compare them with the typical properties of that carrier.

Warden The entity (e.g., a computer program, can also be considered as an attacker of the information hiding method) that monitors the network traffic with intent to spot and/or disrupt/destroy covert communication (see also Prisoner's problem). A *passive* warden passively observes and analyzes network traffic without modifying it. An *active* warden (slightly) distorts a covert communication (so that it still conveys the same overt meaning). A malicious warden is an active warden that alters the messages with impunity and/or that introduces messages into a covert communication by impersonation.

Watermarking Data hiding application in which the main aim is not to enable hidden data exchange between the secret sender and receiver but to conceal information that will supplement the overt data.

INDEX

Information Hiding in Communication Networks: Fundamentals, Mechanisms, Applications, and Countermeasures,
First Edition. Wojciech Mazurczyk, Steffen Wendzel, Sebastian Zander, Amir Houmansadr, and Krzysztof Szczypiorski.
© 2016 by The Institute of Electrical and Electronics Engineers, Inc. Published 2016 by John Wiley & Sons, Inc.

IEEE Press Series on
Information and Communication Networks Security (ICNS)

Series Editor, **Stamatios Kartalopoulos, PhD**

Mission Statement:
This series provides high quality technical books on Information and Communication Networks Security Theory and Technology. The series is interested in the security aspects of all types of communication networks (wireless, wired, optical, quantum, chaotic, hierarchical, non-hierarchical, IP, ad-hoc, cloud, and so on), and in the security of information transported through their nodes and across them. Our security interests are on all levels of the OSI model, from the network physical layer to the application layer, on the node level and end-to-end, and on all levels of mathematical and technical complexity. Books are intended for professionals, researchers, and students, as well as for private, academic and government organizations.